Market Driven
Supply Chains

Market Driven Supply Chains

Amiya K. Chakravarty

College of Business Administration
Northeastern University
Boston, MA, USA

now

the essence of knowledge

Boston – Delft

Published, sold and distributed by:
now Publishers Inc.
PO Box 1024
Hanover, MA 02339
USA
Tel. +1-781-985-4510
www.nowpublishers.com
sales@nowpublishers.com

Outside North America:
now Publishers Inc.
PO Box 179
2600 AD Delft
The Netherlands
Tel. +31-6-51115274

ISBN: 978-1-60198-976-5
© 2010 Amiya K. Chakravarty

Dedication

To my wife, Indira

Loving, invigorating, inspiring, and supporting — always there for me

Preface

The push for efficiency has transformed the supply chains of companies like Toyota and Wal-Mart in the last couple of decades. Through supply chain innovations, these companies have contributed significantly by increasing customer value while enhancing their own profits. The story is not complete however, as customer value cannot be captured through efficiency alone. Simply stated, creation of customer value is the act of satisfying the customer in whatever he/she needs. Some customers want a large variety to choose from or products that are tailor-made, others want rapid response and on-time deliveries without disruption, and still others may want products that they can adapt to their needs through advanced technology. Thus the customer-value space is enormous, providing huge challenges as well as opportunities. To be competitive the companies, in a supply chain, must position themselves appropriately in this space. There are two broad approaches: create capabilities to be competitive in a special niche in this customer-value space, or create responsiveness so that the supply chain can be adapted from one niche to another.

Both the academia and the industrial applications have primarily focused on the first approach: different capabilities for different niches. Only recently does one see emerging themes from industry think-tanks on the need for "demand driven supply network". Responsiveness in supply chains requires "out of the box" thinking. For example, though companies have been creative in coping with demand volatility, they have not paid much attention to shaping the demand to conform to their comfort-zones. Only recently has the airline industry taken a lead role in demand-shaping through dynamic pricing, overbooking, and real-time capacity allocation. The question is which of the supply chain processes — capacity management, production, logistics, procurement, outsourcing, and fulfillment — can be restructured for building responsiveness in the chain, and how.

The recent spate in terrorism and natural disasters has revealed fissures in the management of the humanitarian aspects of supply management. The supply chain objective is not so much the maximization of customer value in the traditional sense but more the minimization of the social cost of disruption. In this case, responsiveness has a much higher return than in a traditional supply chain. Delays do not cause customers to leave; they increase suffering and so increase the social cost.

This book, Market Driven Supply Chains, is conceived to delve in the second approach of competing in niches: creating responsiveness to enable the supply chain to adapt from one niche to another. The focus is on conceptual thinking based on a robust framework. We discuss analytical models where appropriate. The basic framework is discussed at length in Chapter 1 and it outlines how the fit between market drivers and supply chain processes can be assessed in the context of a product or service. The companies can use such fit-assessment to create strategies for the company's evolution in terms of new capabilities. In Chapter 2, we discuss the issues in demand and revenue management for shaping demand. Various concepts of capacity allocation, dynamic pricing, demand allocation, and mass customization are explored. Examples from the airline, fashion goods, and computer industries are discussed. In Chapter 3, the theme is managing capacity for market volatility. Issues such building flexibility, demand allocation between stable and agile suppliers, outsourcing, demand decoupling, capacity trading, and capacity network are explored at length. In Chapter 4, we discuss new product and process configuration for demand volatility, attribute mapping techniques such as the house of quality, fit between supply chains and products, product modules, facility configuration, and re-sequencing, and redesigning processes. In Chapter 5, we emphasize procurement, and the choice of supply network. We discuss issues such as network design, supplier relationship, and contracts. Finally in Chapter 6, we discuss the modalities of response in a supply chain disruption including proactive and reactive approaches. Supply chain fortification against terrorism in a gaming context, and response to natural disasters are discussed. The value of defending a supply chain is explored.

This book is expected to be of interest to both the academics and industrial practitioners, and would be of great value to graduate students in business and engineering. It raises many questions, some provocative, and provides many lead-ins for in depth research. Modeling approaches for new problems are suggested along with discussion of case studies and other examples.

Amiya K. Chakravarty
College of Business Administration
Northeastern University
Boston, MA

Contents

Chapter 1

Markets Drivers and Supply Chains

1.1 Introduction

Consumers are more selective today, and are increasingly concerned with price, delivery, variety, and quality. Companies are employing new strategies that include globalization, using secondary markets (Lee and Whang, 2002), resource sharing, and forming alliances, as they begin to appreciate the importance of supply chains to their bottom lines (Hendricks and Singhal, 2005). While cost-driven supply chain initiatives such as efficient sourcing, lean manufacturing, and third-party logistics have helped to improve productivity, the challenge is to respond appropriately to changing customer preferences, through capability alignment. As an example, preference for increased product variety cannot be satisfied by a highly efficient mass production system. Maintaining alignment with preferences through reconfiguration of processes can be a very demanding in terms of cost, effort, and time. Misalignments, on the other hand, lead to inventory build-up, poor customer service, underutilized facilities, and declining profits. While frequent realignments may cause instability in the supply chain through disruptions, infrequent alignments may hinder adequate response to market opportunities and lead to competitive disadvantage. Development of alignments that retain stability without compromising dynamism would therefore be a strategic objective.

For example, while a flexible system is ideal for responding quickly to changes in demand, it must be built over time through investment in technology, process reengineering, and worker-training. While supply chain capabilities must be developed proactively, the uncertainties in customer-needs set a limit to the degree of proactivity. In the 70s the Allis Chalmers Company invested heavily in flexible processes only to discover a massive downturn in the agriculture industry that bankrupted the company. To fend off the Japanese competition, many American managers invested heavily in flexible processes in the 80s (Jaikumar, 1986). However, unlike the Japanese who used the flexible equipment to increase product variety, the American managers used the new technology for scale economy with disastrous results.

The central issues are how should a company position its capabilities with respect to market opportunities, and how should it evolve with the opportunities. This is a multi dimensional matching problem. For example, the term market-opportunity can only be understood in terms of multiple attributes such as demand size and volatility, product variety, quick response, and quality. Similarly, the attributes defining the supply chain capability are multi faceted: product design, manufacturing capacity, logistics, and procurement. In what follows next, we develop a framework to help establish where a company should position itself in this multi dimensional space. A company may choose to respond to one or more of the market drivers through restructuring the supply chain. Restructuring can be simple as in cost cutting, or extensive with new product launch, adding agility through adaptive processes, and restructuring logistics for quick delivery.

1.2 Market Drivers

Markets may influence operations in a supply chain in multiple ways. These include demand volatility, uncertainties in supplies being delivered on time and in requisite quantities, uncertain yields from processes, innovative technologies that induce changes in supply chain processes, emerging global markets, supply chain disruptions through unanticipated events, and new government regulations related to public health and ethical practices.

The major causes of uncertainty in demand are variable order quantity, large product variety, multiple demand channels, a wide range of service levels, and an uncertain rate of product innovation. Demand variability can also be caused by order cancellations, emergency orders, delayed deliveries, and deliveries to wrong locations. Incorporating new customer segments and/or new markets may induce considerable uncertainty due to information asymmetry. When Dell took its printer business to China it had little information on customer buying habits. Dell had to modify its direct-to-customer model, as the Chinese customers were not comfortable with placing orders online (Ho, 2006). Demand variability can cause the well known bullwhip effect that amplifies demand variance, as the order moves upstream from the customer to the supplier through multiple supply chain tiers. Other causes of variance amplification are anticipated price increases, batching of customer orders, constraints on production capacity, and delivery lead time.

Supply uncertainties, on the other hand, are caused by production capacity, product quality, and fulfillment bottlenecks. Uncertainties in production creep in from poor maintenance and labor turnover that cause shortages. Quality uncertainty stems from product complexity, defective equipment, inadequate worker training, and poor screening. Production processes that are still evolving can cause uncertainty from process modifications. Bottlenecks in fulfillment may cause mismatches between orders and actual deliveries, and increase delays causing missed opportunities.

New technologies such as the Internet, radio frequency identification (RFID), and point-of-sales (POS) have led to a considerable reengineering of business processes. Wal-Mart is one of the companies that took great strides in implementing RFID in its supply chain. Because of the considerable expenses involved, some companies may delay new investment in technology causing the supply chain to be unsynchronized. Therefore, each new wave of technology may require suppliers to retool themselves, and may cause communication problems if partners do not possess synchronized technologies.

The changing business environment may require an overhaul of business processes. Cases in point are sustainable operations through "green" supply chains, organic food, and pollution control (Beamon, 1999); and ethical practices through the Sarbanes-Oxley legislation. These cause businesses to incorporate new processes and/or reengineer current processes, with implications for delays, cost, and revenue.

Today's lean supply chains are more vulnerable than ever to natural and man-made disasters. The potential for disruption comes in many forms, from large-scale natural disasters and terrorist attacks to industrial fires, wide-spread electrical blackouts, and operational challenges such as shipping-port bottlenecks. It is a reality that puts a greater burden on companies to plan for recovery capabilities in the supply chain — proactive as well as reactive.

There are a number of steps a company can take in responding to market uncertainties. It can split demand into stable and variable components; the stable part assigned to a low cost supplier, and the variable part assigned to an agile supplier that can respond to demand variations quickly. Demand shaping is another way to deal with uncertainties. Using incentives, the company may attempt to shift demand to where capacity is available: off peak pricing, advance purchase, and service scheduling. If demand materializes in locations where capacity is not in place (such as global locations), strategic use of partners in an alliance can be useful. In addition, products can be redesigned so that product customization is postponed to a time closer to the actual demand realization.

1.3   Supply Chain

A supply chain is a set of links that connects multiple companies in business relationships. In modeling a supply chain, nodes represent companies, service centers, or production facilities; arcs connecting the nodes represent the flow of goods, information, and money. In a decentralized architecture where companies make decisions based on individual cost structures, coordination in the supply chain becomes crucial (Cachon, 2003). Coordination decisions clearly depend upon the structure (number of nodes and links), and the infrastructure (production, ordering, and fulfillment).

The supply chain can be reorganized to better cope with demand variability in several ways. It can be altered from a hub & spoke architecture to point-to-point (Zapfel and Wasner, 2002). While hub & spoke enhances cost efficiency (major US airlines), the point-to-point architecture is more appropriate for variable customer demand (Amazon.com). A network can link the capacities at supply nodes (locations) in the form of a network, so that capacity shortage at one location can be made up through transshipment from another location with excess capacity. Second, with a high demand variability, the logistics routing needs to be flexible enough to enable frequent deliveries or to deliver to updated delivery plans. Third, in industries with demand uncertainties, such as fashion goods, production can be organized into two runs: a small lot — to learn the demand pattern, and a second larger lot based on better demand information.

In structuring a supply chain, a number of issues need to be considered. First, based on the nature of products and customers, the company must decide the supply structure: hub & spoke, point-to-point, or modular. Note that a modular supply chain would allow fragmented ownership, multiple interchangeable suppliers for key components, and cultural diversity. In fact companies like Dell computers and Cisco Systems thrive on reducing ownership in the chain. By not requiring geographical proximity a modular supply chain can broaden its scope and increase flexibility in its operations, through quick restructuring.

Next, qualified suppliers, manufacturers, distributors, and 3PL partners, to be included in the supply chain need to be identified and organized into tiers (tier 1 closest to the company), hubs, or modules. Note that the bullwhip effect will increase in the number of tiers in the supply chain. Finally, the modus operandi (links, information flow, shipping, payment terms etc.) must be defined for the interlinked nodes. These include which products to target to which destinations in the chain, how to maximize cooperation between customers and producers to minimize waste in the system, how to design flexible contracts among parties, and how to share inventories and profits in a network of suppliers.

The company also needs to establish the mix of channels it would use: inhouse; outsourced; and third party service providers that collect orders, locate suppliers, and deliver orders. For inhouse operations, important consideration would be where to locate the operations, whether they should be decentralized, and whether they should be product or process focused.

1.4   Supply Chain Processes

The capability of a supply chain can be understood in terms of the processes it can perform. These processes can be generalized as customer management, procurement, production, outsourcing, and fulfillment. Customer management process involves assessment of current and future demand by establishing customer preferences through focus groups or customer surveys. It also includes shaping of demand through revenue management initiatives such as capacity allocation, dynamic pricing, and demand allocation.

The procurement process formalizes the acquisition of components, material, and other resources through multiple channels such as online markets, e-auctions, spot markets, and distributors. Decisions involve whether to develop new capabilities of suppliers, whether to use specialized, generic, flexible, or global suppliers, and what types of contracts to negotiate. Negotiating bulk discounts, consignment purchases, lean procurement, invoicing, receiving, and inspections improve procurement efficiency.

Production processes relate to scheduling and sequencing of production and services, batching, and dispatching decisions. Should production be split between two or more partners and, if so, who does what? How does one coordinate the production support functions such as flow of semi finished products, flow of raw material, and the flow of information between partners. Flexibility, automation, and networking of facilities are some of the strategic issues in production and servicing. While flexibility enables a quick response to demand uncertainty, it is not the right option if competing priority is low cost. Thus, a major concern is how much to invest in new technology, when to bring it on board, and how to smoothen its implementation.

The outsourcing processes include make/buy decisions, use of offshore suppliers, and collaboration. Fine (1998) describes Toyota's sourcing policy. Toyota ensures that it has adequate knowledge and capacity for inhouse production of critical components like car engine, and to a lesser degree transmission system. While the resource allocation issue is clear for critical components like engine, it is not so for the less critical components. Toyota completely outsources components such as electrical systems, as they are not critical to Toyota's operations.

The fulfillment process includes the operations of the logistics network, transportation service providers, and expediting. It involves scheduling, status updating, and rerouting. Companies work with logistics partners such as trucking companies. Dynamic capacity allocation is a major issue, as running less than full truck loads can be very expensive. Using a logistics hub with cross docking facility helps improve capacity utilization. Fulfillment effectiveness is one of the competitive advantage of Amazon.com.

With decentralized decision making, managing the relationships between partners is becoming crucial. We may classify relationships as linear or non linear. Linear relationships exist in hierarchies, where information flow is limited to only the adjacent layers (tiers) in the supply chain. Linear relationship is the way most companies conducted business until the advent of the Internet. Now it is possible to communicate with supply partners several tiers upstream or downstream, using online technologies such as RFID and POS. Clearly, not all relationships in a supply chain are strategic. Relationships with strategic partners need careful nurturing. We use the analogy of dating, engaged, and married to describe the portfolio of relationships. As in dating, certain business relationships, though of mutual interest, are casual and rely on continuing assessment of one another. At the "engaged" level, partners develop certain degrees of trust to permit access of each other's databases, whereas, in "married" relationship long term collaboration commitments are made so that decisions of one partner do not hurt others. They form alliances and permit full information visibility, so that one partner may extract data from another using partner's application software.

The operations modes of such relationships can be classified as networking, coordination, cooperation, and collaboration. According to Himmelman (1993), networking facilitates exchange of information; coordination aims at achieving a common purpose that is mutually beneficial by altering information-exchange activities; cooperation facilitates resource sharing for mutual benefit; and collaboration aims at enhancing the effectiveness of a partner by altering activities and sharing resources. Note that while collaboration creates competitive advantage, it may be risky in terms of protection of proprietary information and loss of operational flexibility.

One of the issues in managing relationships is creating connectivity (links) between the tiers. It can be achieved through technology such as electronic data exchange (EDI), Internet, Intranet, and workflow. The type of connectivity chosen must match the nature of relationship. While EDI would work in a point-to-point link, workflow would be necessary for collaboration using hub & spoke.

Contracts replace directives in a decentralized, market driven supply chain. The basic forms of contracts are geared to price and effort, but there could be multiple variations for creating operational flexibility, and capturing different business practices.

Lee (2002) categorizes supply chains into efficient, responsive, risk hedging, and agile, to match products (functional and innovative) with processes (stable and evolving), as in Figure 1.1. Thus, functional products are assigned to efficient or risk hedging supply chains, depending upon process stability; innovative products are assigned to responsive or agile supply chains, again, depending upon process stability.

It follows that an efficient supply chain using lean processes would work well for well established standard consumer products such as TV

	Functional Products	Innovative Products (demand uncertainty)
Stable supply process	*Efficient SC* JIT, lean Logistics efficiency	*Responsive SC* Modular products, supplier hub, postponement
Evolving supply process (supply uncertainty)	*Risk hedging SC* Inventory pooling, Information visibility	*Agile SC* Risk-hedging and responsive

Fig. 1.1 Matching products with supply chain.

and home appliances, where the supply processes have stabilized. This is where WalMart excels. Note that delivering music through CDs has used stable supply processes. But today, with online music delivery, using multiple channels such as itunes, and ipod has induced considerable uncertainties in the supply chain.

Note that functional products that are still new can cause uncertainty, as suppliers experiment with processes to fine tune them. There will be uncertainty about supplier's cost and delivery times, inducing risk in the system. Innovative products such as the GPS for cars (or online communities such as the YouTube and Twitter) face considerable risk in their early development processes. As product design keeps evolving, products are configured in modules and customization is postponed as long as possible. This causes stress in the supply chain; manufacturers may prefer procurement from supply hubs where a wider choice of suppliers is available. Finally, if there are uncertainties in both supply and demand, the supply chain will need to be agile to cope with rapid demand changes. Both Zara and Li & Fung retain considerable agility by creatively using a portfolio of inhouse and outsourced suppliers.

1.5 Linking Market Drivers to Supply Chain Processes

We conceptualize a market driven supply chain as in Figure 1.2, where the fit between the market drivers and supply chain processes is assessed. The strength of fit is depicted by the size of the cross in respective cells, as shown for a fictitious supply chain.

In Figure 1.2, the most effective response to increased demand uncertainty or increased customization would be strengthening the customer management processes such as demand shaping and dynamic pricing, discussed in Chapter 2. The company can also restructure its production processes, but the marginal benefit may not be as high as that from customer management. In the same way an effective response to increased disruptions would be enhanced procurement and fulfillment processes, discussed in chapters 5 and 6. A good example of a misfit between demand drivers and supply chain processes would be outsourcing for an emerging innovative technology. Emerging

Supply Chain Processes

	Customer Mgmt	Procurement	Production	Outsourcing	Fulfillment
Demand uncertainty	x		x	x	x
Supply uncertainty		x	x		x
Customized products	x		x	x	x
Disruptions	x	x	x		x
Quick response	x		x	x	x
Process yield		x	x	x	x
Technology	x	x	x	x	x
Cost		x	x	x	x

Fig. 1.2 Matching market drivers with supply chain.

technologies are knowledge-intensive and they can be developed to provide a new core competency. It may be risky to share it with a partner initially. They also need careful nurturing and can be highly risky. In addition, the outsourcing partner may not be able or willing to contribute much to the success of the technology.

To appreciate the concept of fit between markets and supply chain, we discuss a few examples next. Adidas incorporated mass customization principles in its athletic shoes in 2003 (Seifert, 2003). It designed a special kiosk where customers could walk in to have the contours of their feet recorded electronically. The specifications were fed to manufacturing individually and the finished shoes were delivered in two weeks. The initiative was well received by customers who welcomed the idea of tailored shoes with individual fitting. In the matrix in Figure 1.3, this would impact the market driver called customization with reengineering implications for the production process. Adidas, for various reasons decided to produce these shoes in their existing production lines. Consequences were severe — the new shoes needed to run in batches of one requiring production line to be reconfigured frequently, thus increasing costs tremendously.

Dell's business model, direct-to-customer (DTC) is well known. Dell cut out the intermediaries by letting customers configure their computers online and place orders. However, when they took their printer business to China (Ho, 2006), they experienced different market drivers and had to rethink their model. A large number of the Chinese customers were not computer savvy, and were unable to use the web. A second, more interesting, issue was their suspicion of online payment systems. Because of these problems, Dell had to redesign their supply chain with more brick-and-mortar stores. In the context of Figure 1.2, the market driver customers were concerned about is disruption, and Dell addressed it through a better design of fulfillment.

Another supply chain, often mentioned, is that of WalMart (Chandran and Gupta, 2003). The market driver in this case is cost minimization, and WalMart has reconfigured their supply chain. In the context of Figure 1.2, they have automated the ordering and fulfillment processes, incorporated cost saving technologies such as RFID and web based ordering, and have worked with a preferred list of pre qualified suppliers. They have reengineered the procurement process with consignment orders and CPFR (collaborative planning, forecasting, and replenishment) to minimize inventories in the system.

WebMD employs extensive outsourcing in its supply chain to acquire content for a wide variety of customer inquiries (Coughlan et al., 2001). It uses an elaborate payment system for its content partners, based on the actual usage of content by customers. Syndication services such as Reuters use targeted-delivery of content to customers based on content sign-up "sheets". Monthly (or weekly) charges are based on what customers sign-up. The appropriate market driver in Figure 1.2 would be the technology of content creation and the matching supply chain process would be content delivery (fulfillment).

Amazon.com provides customization by sensing customers' reading habits (Linden et al., 2003). It outsources the task of sensing the patterns in reading preferences to partners who use advanced datamining techniques. The market driver is demand uncertainty and

Amazon relies on minimizing this uncertainty through early detection of demand, and pre positioning the suppliers.

1.6   Restructuring the Supply Chain

Consider an airline that is considering adding a premium service. The market drivers and their corresponding supply chain processes for premium and average services are listed in Figure 1.3.

Let us assume that average service is what the company is focused at. It follows from Figure 1.3, that to satisfy the market drivers for premium service the airline will need to develop new capabilities: offer customized food service, offer flat bed, offer private space in cubicles, facilitate itinerary changes with partner airlines etc. Each of the new capabilities may require new knowledge and may involve creation of new procedures and training of employees. It certainly would need considerable new investment. It would be a mistake to try to offer a "premium service" with a slight retooling of current standardized processes, as it would cause a misfit. Some airlines may do a better job in creating the new capability and thus differentiate themselves from the rest. Of course, even if one could create the new capabilities, additional revenue from the new service must justify the new investments.

It can happen that the market drivers shift from low price and "acceptable" service to customization. Using a framework as in Figure 1.3, the company can identify the new supply chain processes that would be required in the new market environment. If the airline cannot build new capabilities quickly, it may consider aligning with other partners who can. It can then act as an intermediary and transfer customers seeking premium service to the partners.

	Premium Service	Average Service
Market Drivers	Customization, additional services, comfort, convenience	Price, acceptable service
Supply chain processes	Creating and/or procuring flexible services, facilitate customer driven changes with partner airlines	Standardized service processes, procure food services from low cost vendors

Fig. 1.3 Matching market drivers with processes.

Clearly the constructs in Figures 1.2 and 1.3 provide a robust framework for assessing restructuring decisions in a supply chain. Next, in chapters 2 to 6 we consider in detail the capabilities of customer management, supply chain capacity restructuring, product and process configuration, procurement and supplier network. Finally we explore the impact of a new market driver — supply chain disruption.

Chapter 2

Demand and Revenue Management

2.1 Introduction

Customers, invariably, want new or improved products, and the rapid technological innovation is enabling it. While the demand for functional products such as toothpaste is predictable, predicting customer demand for fashion goods is not easy. As a consequence, manufacturers and retailers find themselves holding huge inventories at high cost. Lee (2003) points out that the retail chain in the United States has $1.1 trillion in inventory on average (estimated by the United States Commerce Department). Of this, $400 billion are at retail locations, $290 billion at wholesalers, and $450 billion with manufacturers.

As a result, companies invest in flexibility or balance inventories against shortages — the so called coping strategy (Gerwin, 1993). In contrast, the risk reduction strategy aims at altering customer's requirement to fit with the capabilities of the supply chain. It does so by influencing the customer's purchase behavior. Airlines influence behavior in multiple ways: controlling the allocation of capacity (number of seats) in a fare class, overbooking, and through dynamic pricing. Dell Computer promotes product-configurations that include more of its surplus components, on its website. Between the two extremes of coping and demand-shaping, other hybrid strategies are possible. Some examples are: allocating variable demand to a portfolio of agile and focused suppliers to diversify risk, consolidating demand for a wider servicing window, and permitting flexibility in delivery arrangement. Finally, structuring appropriate contracts help induce risk-sharing with partners. While the airline and Dell Computer's decisions are based on appropriate pricing and/or capacity allocation (known as revenue management), the hybrid strategies may need restructuring of the supply chain.

2.2   Revenue Management

Talluri and van Ryzin (2004) outline two primary strategies for revenue management: capacity control, and price control. Capacity control is based on allocating capacity while the price is fixed, overbooking, and through service customization. An airline can dynamically increase or decrease the number of seats it sells in a fare class. It uses over-booking to maximize seat utilization by minimizing the risk of no-shows. It creates a hedge against no-shows, but incurs the risk of having to buy-back reserved seats through auction (when most customers show up). Capacity reservation is another way of moving demand to "capacity categories", and it acts like a financial option. Customers may reserve capacity ex ante for a fee. They use the reserved capacity first and, if needed, the unreserved capacity at a higher price. They forfeit the unused reserved capacity. Discount pricing for early commitment provides an opportunity to fill a part of the capacity with certainty.

Price control strategies include dynamic pricing as in auctions, and price discrimination in customer segments. Product markdowns help sell off excess inventory (overstock.com). However, if markdowns become predictable, customers may wait until the period-end thereby reducing company's revenues. Promotions through advertisements or other marketing techniques create additional demand to fill unused capacity (Bursa, 2009). Utility companies practice peak-load pricing to smooth the high demand for telephone service during peak hours. Price discrimination relies on being able to create customer segmentation based on willingness to pay. For example, negotiation may result in different prices for urban and rural customers buying the same service. Similarly, special orders such as services that are made-to-order can be created for customers who are willing to pay more.

2.2.1 Capacity Allocation

Capacity based revenue management attempts to maximize revenue with given prices (Mcgill and Van Ryzin, 1999). The key is customer segmentation so that different prices may be charged to different segments, based on price sensitivity and the need for convenience and comfort. The airline industry uses fare classes Y, M, B, and Q to channel customers into segments. The number of seats assigned to a segment would be a function of the type of fare-class, and the demand pattern.

Consider a seat inventory of y and two customer segments with fixed prices such that $p_1 < p_2$. Assume that demand in the two segments are $D(p_i) + \xi_i, i = 1, 2$, where ξ_i is the random component of demand. Assume that a customer wishes to buy a seat in segment 1 and that this demand cannot be backlogged. The decision would be simple if the airline could not sell more than $y - 1$ seats in segment 2. However, as there is a non zero probability of selling the yth seat in segment 2, the airline will sell the seat to segment 1 only if

$$p_1 > p_2 \int_{\xi_2 \geq y - D(p_2)} f(\xi_2) d\xi_2 = p_2 \{1 - F(y - D(p_2))\}.$$

That is,

$$y > D(p_2) + F^{-1}\left(1 - \frac{p_1}{p_2}\right) = \hat{y}$$

Thus, the seat inventory must exceed a threshold, for a seat in the lower price segment to be sold. It follows that for given segment-prices, the seat inventory threshold can be determined uniquely. Thus, a simple decision rule of the type $y > \hat{y}$ is sufficient for real time implementation. Clearly, the revenue from segment 2, $p_2\{1 - F(y - D(p_2))\}$, increases if y is reduced. Also, note from the condition on y that the price of a seat in segment 1 (low-price segment) p_1 increases if the airline's seat inventory threshold $\hat{y}$ depletes — a common practice in airline pricing.

Leakage between demand segments can be a problem in allocating seats to the segments. To avoid it, companies create fences between segments. In the airline industry the fence may take the form of a high penalty for making changes, advance-purchase, or Saturday-night-stay.

2.2.2 Dynamic Pricing

Determining prices based on how customers value products and services, at any point in time, is known as dynamic pricing (Bitran and Caldentey, 2003). It has become a feasible strategy in online transactions, as sellers and buyers can make rapid adjustments to prices at which products are bought and sold. For example, sellers offer special deals, targeted to individual customers, and buyers can quickly scan to compare prices and services. It is now possible to offer an item at different prices, vary price over time, and charge different prices to different consumers for the same product. A well known example is the patented drug that fetches a higher price in US than in Canada. Other examples are group pricing, and differential telecom pricing for businesses and households.

Dynamic pricing methods can be categorized into two groups: posted price and price discovery. Posted prices are determined by the seller and hence not flexible. However, they could be dynamic in the sense that the seller could change posted prices periodically depending on demand, supply availability, and the time of sale. In price discovery mechanisms, prices are determined through a bidding process such as an auction.

2.2.2.1 Demand shaping

Demand shaping is the practice of increasing or decreasing demand by suppliers, to match demand with their existing capacity (Jinhong and Shugan, 2001). It is a major issue in a robust and growing economy. On the surface this does not appear to be a problem. However, the way it usually works is that early customers get what they want and customers who delay their decisions have their requests denied. Because some customers in this later group would be willing to pay a higher price, there is a loss of opportunity as money is left on the table. Obviously if all customers arrive simultaneously, the supplier could cherry pick high-paying customers and her profit would be maximized. A generic approach would be to increase price for all so that demand is reduced to match available capacity. This is a variation of the capacity allocation problem discussed earlier; now the decision variable is price. By including the randomness in demand, the decision becomes complex.

Assume the demand to be $D(p) + \xi$, where $D(p)(D'(p) < 0$ and $D''(p) < 0$) is the known component of demand (at price p), and ξ the random component with mean zero. It can be shown that without a capacity constraint supplier's profit is written as

$$\pi = pE_\xi\{D(p) + \xi\}$$

As the profit function is concave in p, the profit maximizing price is expressed as

$$D(p^*) + p^*D'(p^*) = 0 \qquad (2.1)$$

The expected value of used capacity would be $D(p^*)$, and the used capacity will have a wide range of variation from its expected value.

If available capacity is limited at Q, supplier's expected revenue would be written as

$$\pi = pE_\xi\{\min(D(p) + \xi, Q)\}$$

As π is concave in p, using the first order condition, optimal p is expressed as

$$\{D(p^*) + p^*D'(p^*) - Q\}F(Q - D(p^*))$$
$$+Q + \int_{\xi \leq Q - D(p^*)} \xi f(\xi)d\xi = 0 \qquad (2.2)$$

As $D'(p) < 0$ and $D(p) + pD'(p)$ decreases in p, it follows that p^* in (2.2) will be larger than that in (2.1). Expression (2.2) is simplified to $p^*D'(p^*)F(Q - D(p^*)) + Q = \int_{\xi \leq Q - D(p^*)} F(\xi)d\xi$.

If the known demand component is linear so that $D(p) = A - p$, we can solve for optimal p as

$$p^*F(Q - A + p^*) + \int_{\xi \leq Q - A + p^*} F(\xi)d\xi = Q$$

Willingness to Pay (WTP)

We can simplify the demand shaping problem by using the willingness to pay (WTP), to define the demand function discussed below. Assuming willingness to pay (WTP) to be a random variable with distribution $G(\cdot)$, the probability that customers will be willing to pay p or more will be $1 - G(p)$. Hence, retailer's revenue will be (assuming market size of 1, and $0 \leq Q \leq 1$),

$$R = p\{1 - G(p)\}$$

$$\text{Subject to } 1 - G(p) \leq Q$$

It can be verified that R is unimodal in p (assuming IGFR $G(\cdot)$). Therefore, R is maximized at

$$1 - G(p^o) - p^o g(p^o) = 0$$

Also, the price at which all the inventory is sold out is expressed as

$$p(Q) = G^{-1}(1 - Q)$$

Thus, the optimal price will be

$$p^* = \min\{p^o, p(Q)\}$$

If the demand possesses a random component ξ, so that demand equals $1 - G(p) + \xi$. Revenue is now written as

$$R = \int_{\xi \leq Q - 1 + G(p)} p\{1 - G(p) + \xi\} f(\xi) d(\xi) + \int_{\xi \geq Q - 1 + G(p)} p Q f(\xi) d(\xi)$$

To maximize R, we have

$$\frac{\partial R}{\partial p} = \{1 - G(p) - pg(p)\} F(Q - 1 + G(p)) + Q\{1 - F(Q - 1 + G(p))\}$$

Assuming IGFR $F(\cdot)$, it can be shown that R is unimodal with a unique maximum in p. Therefore, from the first order condition, we solve for p as

$$\{Q - 1 + G(p) + pg(p)\} F(Q - 1 + G(p)) = Q$$

It can be verified that the randomness in demand increases the optimal price.

2.2.3   Surplus Capacity or Inventory

To get rid of surplus components, Dell alters its web site almost every day to promote product configurations that include the surplus components. Cutting prices on these product configurations leads to increase in demand so that more of the surplus components are absorbed. As before, assume $D(p)$ to be the demand of the product. Assume Q to be the amount of surplus components and c its unit cost. Thus, Dell saves an amount c per unit of surplus component used. Dell's revenue, inclusive of the cost-saving on components, is written as

$$\pi = pE_\xi\{D(p) + \xi\} + cE_\xi\{\min(Q, D(p) + \xi)\}$$

As π is concave in p, using the first order condition the profit maximizing p is expressed as

$$D(p^*) + p^*D'(p^*) + cF(Q - D(p^*))D'(p^*) = 0, \qquad (2.3)$$

As $D'(p) < 0$ and $D(p) + pD'(p)$ decreases in p, it follows that p^* in (2.3) will be smaller than that in (2.1). That is, Dell benefits by reducing the optimal price on products that can absorb surplus components.

With $D(p) = A - p$, we would have $2p + cF(Q - A + p) = A$. Note that p decreases in both c and Q as expected; the rate of decrease in c exceeds that in Q.

2.2.4   Capacity Allocation with Pricing

Consider a simple example. A product can be sold at unit price p_1 or in a bundle of six at unit price p_2 per bundle. Assume that the company has Q units of the product in inventory and that demand for the two categories are $D(p_i), i = 1, 2$. The firm needs to know how to divide the inventory between the two categories and how much to charge for each category. First, ignoring the random components of demand, we express the revenue function as

$$R = p_1 D(p_1) + p_2 D(p_2)$$

$$\text{Subject to } D(p_1) + 6D(p_2) = Q$$

Assuming linear demand function, $D(p_i) = A_i - p_i, i = 1, 2$, and using Lagrange' multiplier λ, it can be shown that for the optimal solution

$$\lambda = \frac{A_1 + 6A_2 - 2Q}{37}, \quad p_1 = \frac{19A_1 + 3A_2}{37} - \frac{Q}{37},$$

$$p_2 = \frac{6A_1 + 73A_2}{74} - \frac{6Q}{37}$$

$$D(p_1) = \frac{18A_1 - 3A_2}{37} + \frac{Q}{37}, \quad D(p_2) = \frac{A_2 - 6A_1}{74} + \frac{6Q}{37}$$

The assumption that demands in the two segments are independent may not hold, as p_1/p_2 must be sufficiently low for customers to be willing to buy in singles. Therefore, we need the condition, $ap_1 < p_2 < 6p_1$, $1 < a < 6$, to ensure that neither segment is empty. This simplifies to the pair of constraints involving A_i; $A_2 < 6A_1$ and $Q < \frac{(73-6a)A_2-(38a-6)A_1}{2(6-a)}$. These two expressions together express the nature of dependence between the two demand functions that make the two-segment solution feasible.

Weatherford (1997) presents a formulation of the simultaneous pricing/allocation decision that assumes normally distributed demands. For the general case, with stochastic demand, we have

$$R = p_1 E_{\xi_1}\{\min(Q, D(p_1) + \xi_1)\}$$

$$+ p_2 E_{\xi_2}\{Q - D(p_1) - \xi_1, D(p_2) + \xi_2\}, \qquad (2.4)$$

The first order conditions of (2.4) are obtained from the following two equations.

$$\frac{\partial R}{\partial p_1} = \int_{\xi_1 \leq Q - D(p_1)} \{D(p_1) + p_1 D'(p_1) + \xi_1\} f_1(\xi_1) d\xi_1$$

$$+ Q \int_{\xi_1 \geq Q - D(p_1)} f_1(\xi_1) d\xi_1$$

$$\frac{\partial R}{\partial p_2} = \int_{\xi_1 \leq Q - D(p_1)} \left\{ \begin{array}{c} \int_{\xi_2 \leq Q - \xi_1 + D(p_1) - D(p_2)} \{D(p_2) \\ + p_2 D'(p_2) + \xi_2\} f_2(\xi_2) d\xi_2 \\ + \int_{\xi_2 \geq Q - \xi_1 - D(p_1) - D(p_2)} \{Q - \xi_1 - D(p_1)\} f_2(\xi_2) d\xi_2 \end{array} \right\} f_1(\xi_1) d\xi_1$$

As we see above, the first order conditions become unwieldy even with linear demand functions. This can only be solved numerically.

2.2.5 Contracts

Buyers and suppliers use contractual agreements to reduce the negative impact of variance through behavior modification. Contracts can be designed in different ways involving price, effort, and risk sharing. In a wholesale price contract, where supplier sets price and retailer determines purchase quantity, retailer bears all the risk of demand variability (resulting in unsold goods). A buy-back contract where the supplier agrees to buy back unsold items at a reduced price, shifts some of the risk to the supplier.

An alternative for the retailer is to split the order into two sequential lots. The first lot is a relatively small order, and the retailer uses this to sense customer-response. If the response is positive, retailer places a second much larger order; if not, she discontinues the product. Zara (fashion retailer) has implemented this idea to perfection. They introduce new products in rapid succession and, as demand forecast for new products is not meaningful, they use 2-period split order to sense demand. It does not have the same effect as demand splitting, as demand-update may modify the demand. It works well for Zara, as it is its own supplier for most product lines. For a supplier, supplying to Zara, it would off course want bounds to be placed on the divergence between the first and second order.

Hewlett Packard has used a "portfolio" contract where the customer commits to buy a fixed number of units at a specified price. For this contract, the customer bears all risk. Customers may place a second order with variable quantity for which the supplier may charge a higher price. There is a commitment to give a certain percentage of business to the supplier. HP's contract is similar to demand splitting discussed earlier, but now we have only a single supplier and there is a business commitment to the supplier over and above the fixed quantity contract. In other variations, the variable quantity contract may be replaced by profit sharing contract. A fuller discussion of contracts is included in Chapter 5.

2.3 Demand Management

Next, we consider demand management issues affected by order placement decisions, in addition to pricing decisions (Benjaafar et al., 2004).

2.3.1   Demand Allocation to Match Capacity Profile

Clearly, demand variability can be better managed if the unstable component of demand can be isolated. For example, if demand of a company's product varies between 500 and 800 units per month, it would be risk free to produce the first 500 units. A production quantity, exceeding 500 but not 800 units, will involve both upside and downside risks. As a policy, it would make sense to choose a stable production option for the first Q units of demand, and an agile option for the units of demand exceeding Q. The case in point is Griffin Manufacturing in the US (Warburton and Stratton, 2002). Griffin's customer (a large retailer) planned to move work to Honduras to take advantage of low cost production. The Honduran supplier's production was based on economy of scale, and it could not react to changes in demand quickly. This provided Griffin a niche and it reengineered its process to become an agile supplier, as shown in Figure 2.1. Clearly, splitting of demand created a new competitive environment.

The retailer could now allocate the first Q units of production to the Honduran supplier at a low unit price, and a variable amount based on actual demand, to Griffin at a higher unit price. Because of the variability in order size, Griffin's decision on production capacity was not risk free. Thus, three decisions needed to be made — Honduran order size Q (by retailer), wholesale price w (by Griffin), and production capacity R (by Griffin). We analyze how the parties may make such decisions that remain in equilibrium with respect to one another.

Assume two suppliers: a low-cost offshore (supplier 1), and an agile on-shore (supplier 2). Assume demand to be a random variable ξ, distribution $F(\cdot)$. The retailer structures two contracts: supplier 1 is

	Stable demand	Volatile demand
Agile supplier		Reactive capacity (Griffin)
Low cost supplier	Speculative capacity (Honduras)	

Fig. 2.1 Demand allocation.

contracted to produce quantity Q, supplier 2 is contracted to produce $\xi - Q$ if $\xi > Q$, or R if $R < \xi - Q$, where R is the capacity of supplier 2. The retailer decides Q. The suppliers decide their respective wholesale prices, w_1 and w_2, and supplier 2 decides his capacity R. The retailer's sale price is p. Suppliers' unit prices are c_1 and c_2, respectively. First, the retailer determines Q for given w_1, w_2, and R. The suppliers use the response in Q to determine w_1, w_2, and R.

The expected profits of the three parties are expressed as follows:

$$\text{Supplier 1}: \pi_1 = (w_1 - c_1)Q$$

$$\text{Supplier 2}: \pi_2 = w_2 \left\{ \int_{Q \leq \xi \leq Q+R} (\xi - Q) f(\xi) \right. $$
$$\left. + R \int_{\xi \geq Q+R} f(\xi) \right\} - c_2 R$$

$$\text{Retailer}: \pi_R = \left| \begin{array}{l} p \left\{ \int_{\xi \leq Q+R} \xi f(\xi) + (Q+R) \int_{\xi \geq Q+R} f(\xi) \right\} \\ - w_2 \left\{ \int_{Q \leq \xi \leq Q+R} (\xi - Q) f(\xi) \right. \\ \left. + R \int_{\xi \geq Q+R} f(\xi) \right\} - w_1 Q \end{array} \right.$$

As $p \geq w_2$, it can be verified that the retailer's profit is concave in Q. Hence, from the first order condition Q^* is determined as,

$$(p - w_2) F(Q^* + R) + w_2 F(Q^*) = p - w_1 \qquad (2.5)$$

Supplier maximizes his profit in w_1, subject to (2.5) as a constraint. It is shown in the appendix that π_1 is unimodal in w_1. Therefore from the first order Equation (2.6), we solve for Q^* and hence for w_1 in (2.5).

$$(p - w_2)\{1 - F(Q^* + R) - Q^* f(Q^* + R)\}$$
$$+ w_2 \{1 - F(Q^*) - Q^* f(Q^*)\} = c_1 \qquad (2.6)$$

Next, supplier 2 maximizes his profit in w_2 and R. It is shown in the appendix that π_2 is unimodal in R and w_2. Hence, from the first order conditions in R and w_2, we have

$$\frac{(p_j - w_{2j})F(Q_j)f(Q_j^* + R_j) + w_{2j}F(Q_j + R_j)f(Q_j^*)}{(p_j - w_{2j})f(Q_j^* + R_j) + w_{2j}f(Q_j^*)} = 1 - \frac{c_{2j}}{w_{2j}}$$

$$(2.7)$$

Finally, supplier 2 maximizes his profit in w_2. As shown in the appendix π_2 is unimodal in w_2.

Therefore from the first order condition we have,

$$\left. \begin{array}{c} \int_{Q_j^* \leq \xi_j \leq Q_j^* + R_j} (\xi_j - Q_j^*)f(\xi_j) + R_j \int_{\xi_j \geq Q_j^* + R_j} f(\xi_j) \\ \\ -\frac{w_{2j}\{F(Q_j^* + R_j) - F(Q_j^*)\}\{F(Q_j^* + R_j) - F(Q_j^*)\}}{(p_j - w_{2j})f(Q_j^* + R_j) + w_{2j}f(Q_j^*)} \end{array} \right| = 0,$$

$$(2.8)$$

Optimal Q, R, and w can be obtained from equations (2.6), (2.7), and (2.8).

2.3.2   Demand Consolidation (risk pooling)

Consolidation of demand from two or more sources can lead to better utilization of capacity (Gerchak and Q, 2002). The intuition is that with no consolidation, if demand in one location is higher than expected it may run out of inventory even if there are extra units available in the other location where demand is lower than expected. Thus, if the inventories for the two locations are consolidated, high demand at one location can be offset by low demand at the other location. If demand at both locations is high, there may be shortages even with consolidation. Clearly, consolidating demands that are negatively correlated, leads to greater savings.

Consolidation can occur in three different ways: by location, by time periods, and by products. Consolidation by location has resulted in companies setting up regional distribution centers, and serving multiple cities from the same warehouse. Staples has used centralized US distribution network (Gourley, 1997). Disney Stores uses a dedicated Central DC (CDC) in Memphis to supply more than 100000 types of products to 360 stores (Jedd, 1996). Benetton uses one CDC in Ponzano, Italy, to serve over 6000 stores in 83 countries (Dapiran, 1992).

Aggregating demand across several periods allows cost-effective smoothing of production and consolidation of orders. For example, it can lead to a wider service window that permits aggregation of customer demand, reducing *variability* of orders. Another example is flexible delivery commitments that enable reliable and cost effective demand fulfillment. Cisco Systems often uses a 21-day delivery window despite the fact that many orders can be filled in 10 to 15 days. It achieves almost perfect order fulfillment with more time to fill each order. As delivery-window is not a competitive differentiator for Cisco Systems, increasing it from 10 to 21 days does not hurt them. Amazon.com consolidates orders to optimize picking, packing and shipping operations by quoting a flexible service window of five to nine days rather than a rigid one. Dell consolidates parcel packages for each of its hubs and uses zone-skipping freight rates in conjunction, to reduce logistics cost.

Consolidation by products is possible if they are complements. In the aviation industry it has led to the formation of horizontal alliances and code sharing. Code sharing allows an airline to sell tickets on a partner airline on routes where the partner provides connecting services. There are many examples of code sharing alliances that include United and Lufthansa, JAL and British Air, and JAL and United.

Retailers in food and software industries frequently offer bundled products (Stremersch and Tellis, 2002). They may bundle two products with relative ease without altering the identity of individual products, and sell them together as a "pack". McDonald sells drinks and burgers together at a discount. Amazon.com offers to sell books in a bundle with another book at a discount price. We next analyze how retailers bundles may alter when supplier's pricing decisions are included in bundling decisions.

Consider two suppliers supplying one product each that can be bundled by the retailer. Suppliers' wholesale prices are w_1 and w_2 and retailer's price is $p = w_1 + w_2 + m$, where m is the margin. Suppliers' variable costs are c_1 and c_2. The number of customers purchasing the bundle will depend on the customers' utility or willingness to pay (WTP). With WTP as a random variable with distribution $F(\cdot)$, the probability of buying the bundle at price p will be $1 - F(p)$. The supplier maximizes his profit

$$\pi_S(w_i) = (w_i - c_i)\left\{1 - F\left(\sum_{j=1,2} w_j + m\right)\right\}, \quad i = 1, 2$$

With IGFR $F(\cdot)$, it can be shown that $\pi(w_i)$ will be unimodal in w_i. Hence from the first order condition, we have

$$1 - F\left(\sum_{j=1,2} w_j + m\right) - (w_i - c_i)f\left(\sum_{j=1,2} w_j + m\right) = 0, \quad i = 1, 2,$$

$$(2.9)$$

The retailer will maximize her profit

$$\pi_R(m) = m\left\{1 - F\left(\sum_{j=1,2} w_j + m\right)\right\},$$

As $\pi_R(m)$ is concave in m, we have from the first order condition for optimal m,

$$1 - F\left(\sum_{j=1,2} w_j + m\right) - mf\left(\sum_{j=1,2} w_j + m\right) = 0, \quad (2.10)$$

We can now solve (2.9) and (2.10) for m and $w_i, i = 1, 2$. It follows easily that

$$m = w_i - c_i, \quad i = 1, 2$$

That is, the suppliers' margins will equal the retailer's margin m. Using this property we can solve for m as

$$1 - F\left(3m + \sum_{j=1,2} c_j\right) - mf\left(3m + \sum_{j=1,2} c_j\right) = 0, \quad (2.11)$$

Chakravarty et al. (2010) show that retailers would participate in bundling products, only if product demands are highly (negatively) correlated. However, they point out (as shown in the appendix) that bundling with the constraint on the retailer's reservation price coordinates the supply chain independently of correlation.

2.3.3 Demand Postponement

Inadequate inventory or insufficient number of servers causes delays to customers, resulting in negative utility. Therefore, a popular way to cope with demand volatility is to provide for buffer inventory or excess service capacity, which can be expensive. Examples come mostly from the service industry: waiting lines at bank-teller windows, hair salons, and check-out counters. One way to avoid waiting is to schedule the services and then have customers fit in, as in scheduled airline flights and appointments at a doctor's clinic. The underlying assumption is that the negative utility per unit of delay is the same for all customers. The service providers may compete by increasing service capacity to decrease delays. Note that the notion of appointments — barring rain-checks — is hard to implement in retail and manufacturing industries.

A more effective way to influence demand would be to pay the customer for waiting. This payment could take the form of a price discount that would be proportional to the length of delay. The most relevant work is that by Li (1992), who considers a model where customers choose to purchase from the firm that has the earliest delivery time. Note that in general the supplier gains from reduced inventory or capacity, moderated by the reduction in its unit revenue. The customer, on the other hand trades off the savings from lower unit cost with the negative utility of waiting. Assume $p(t)$ and $w(t) + \xi$ (where ξ is a random variable with distribution $F(\cdot)$) to be the unit price and waiting-disutility, respectively, corresponding to a waiting time t. Therefore, customer's total disutility is

$$u(t) = p(t) + w(t) + \xi$$

$$p'(t) < 0, \quad p''(t) > 0, \quad w'(t) > 0, \quad w''(t) > 0$$

Assume that customer buys the product if $u(t) \leq r$. Then, we can express the probability of making a purchase as

$$\Pr(u(t) \leq r) = F(r - p(t) - w(t))$$

Sales, with M as the market size, is written as

$$S(t) = M\{F(r - p(t) - w(t))\}$$

As $p(t)$ is convex decreasing and $w(t)$ is convex increasing, $r - p(t) - w(t)$ is concave with a unique maximum. Therefore, $F(r - p(t) - w(t))$ and $S(t)$ are also concave with unique maximum.

Denoting the supplier's inventory as $I(t)$ and inventory carrying cost as h, supplier's surplus is expressed as

$$\pi(t) = Mp(t)\{F(r - p(t) - w(t))\} - I(t)h$$

$$\frac{\partial \pi(t)}{\partial t} = M[p'(t)F(r - p(t) - w(t))$$

$$+ p(t)F'(r - p(t) - w(t))] - I'(t)h$$

At $t = 0$, the first term inside the square parenthesis is zero because $Mp(t)\{F(r - p(t) - w(t))\}$ is concave with a turning point, and the second term is positive. Hence, $\frac{\partial \pi(t)}{\partial t} > 0$, as $I'(t) < 0$. At an intermediate point where $F'(r - p(t) - w(t)) = 0$, we have $\frac{\partial \pi(t)}{\partial t} < 0$. At a large value of t, both terms inside the parenthesis are negative. Therefore, $\pi(t)$ is unimodal with a unique maximum. Hence, from the first order condition,

$$p'(t)F(r - p(t) - w(t)) + p(t)F'(r - p(t) - w(t)) = I'(t)h/M$$

Sun et al. (2008) discuss how an online firm with no inventory can compete by discounting the unit price. Assuming p_i and t_i as the unit price and waiting time for firm i's product, they establish firm i's market share as

$$m_i = \Pr\left[\{p_i + \lambda(t_i + v\beta(t_i > 0))\} \le \min_{j \neq i}\{p_j + \lambda(t_j + v\beta(t_j > 0))\}\right]$$

where λ is a random variable capturing market heterogeneity so that λt is the time-based sensitivity to waiting and λv is the fixed component of the disutility due to the hassle or frustration caused by a backorder, and $\beta(\cdot)$ is the indicator function.

2.4 Appendix

1. It follows from supplier 1's profit function $\pi_1 = (w_1 - c_1)Q^*$ that

$$\frac{\partial \pi_1}{\partial w_1} = (w_1 - c_1)\frac{\partial Q^*}{\partial w_1} + Q^*$$

From (2.5),

$$\frac{\partial Q^*}{\partial w_1} = -\frac{1}{(p - w_2)f(Q^* + R) + w_2 f(Q^*)}$$

Substituting $w_1 = (p - w_2)\{1 - F(Q^* + R)\} + w_2\{1 - F(Q^*)\}$ and $\frac{\partial Q^*}{\partial w_1}$ in $\frac{\partial \pi_1}{\partial w_1}$, it can be verified that

$$\frac{\partial \pi_1}{\partial w_1} = \frac{\begin{array}{c}-(p - w_2)\{1 - F(Q^* + R) - Q^* f(Q^* + R)\}\\ - w_2\{1 - F(Q^*) - Q^* f(Q^*)\} + c_1\end{array}}{(p - w_2)f(Q^* + R) + w_2 f(Q^*)}$$

As Q^* decreases in w_1 and as $F(x)$ possesses IGFR property, we can establish from the above expression that π_1 is unimodal in w_1.

2. It follows from supplier 2's profit equation,

$$\frac{\partial \pi_2}{\partial R} = w_2\{1 - F(Q + R)\}$$
$$+ \frac{w_2\{F(Q + R) - F(Q)\}(p - w_2)f(Q^* + R)}{(p - w_2)f(Q^* + R) + w_2 f(Q^*)} - c_2,$$

It follows from (2.3) that $\lim_{R \to \infty} F(Q^*) = 1 - \frac{w_1}{p}$, implying $\lim_{R \to \infty} Q^* > 0$. As $\lim_{R \to \infty} f(Q^* + R) = 0$, and that $f(Q^*) > 0$ if $Q^* > 0$, it is clear that $\lim_{R \to \infty} \frac{(p - w_2)f(Q^* + R)}{(p - w_2)f(Q^* + R) + w_2 f(Q^*)} = 0$. Thus, $\lim_{R \to \infty} \frac{\partial \pi_2}{\partial R} = -c_2 < 0$. Also, $\lim_{R \to 0} \frac{\partial \pi_2}{\partial R} = w_2\{1 - F(Q)\} - c_2 = \frac{w_1 w_2}{p} - c_2 > 0$, if c_2 is sufficiently small. Hence π_2 is unimodal in R.

3. It follows from supplier 2's profit that

$$\frac{\partial \pi_2}{\partial w_2} = \left| \frac{\int_{Q^* \leq \xi \leq Q^* + R}(\xi - Q^*)f(\xi) + R\int_{\xi \geq Q^* + R}f(\xi)}{-\frac{w_2\{F(Q^* + R) - F(Q^*)\}\{F(Q^* + R) - F(Q^*)\}}{(p - w_2)f(Q^* + R) + w_2 f(Q^*)}} \right.$$

Clearly, $\lim_{w_2 \to 0} \frac{\partial \pi_2}{\partial w_2} > 0$. Also, $\lim_{w_2 \to p}\{\int_{Q \leq \xi \leq Q + R}(\xi - Q)f(\xi) + R\int_{\xi \geq Q + R}f(\xi)\} = 0$ because the buyer will not purchase from supplier 2 if $w_2 = p$. Hence, $\lim_{w_2 \to p} \frac{\partial \pi_2}{\partial w_2} < 0$. Hence, π_2 is unimodal in w_2.

4. Bundling products in a supply chain, with retailer's reservation value, coordinates the supply chain. This can be established by noting that the retailer would participate in bundling if her profit $m\{1 - G(w_1 + w_2 + m)\}$ is at least as large as her reservation value R ($R \geq$ profit from the unbundled scenario). Therefore, supplier i maximizes $(w_i - c_i)\{1 - G(\sum_i w_i + m)\}$ with $\sum_i w_i \leq G^{-1}(1 - \frac{R}{m}) - m$. Considering the constrained optimization with Lagrangean multiplier λ, and the IGFR property, it follows from the first order conditions in w_i and λ that $\lambda = 1 - G(\sum_i w_i + m) - (w_i - c_i)g(\sum_i w_i + m)$ and $\sum_i w_i + m = G^{-1}(1 - \frac{R}{m})$. Eliminating $\sum_i w_i + m$, we have $w_i - c_i = \frac{R/m - \lambda}{g(G^{-1}(1 - R/m))}$. Solving for λ and w_i, $2w_i = G^{-1}(1 - \frac{R}{m}) - m + c_i - c_{3-i}$, $\lambda = \frac{R}{m} - \frac{1}{2}\{G^{-1}(1 - \frac{R}{m}) - (m + \sum_i c_i)\}g(G^{-1}(1 - \frac{R}{m}))$, and $m_i = w_i - c_i = \frac{1}{2}\{G^{-1}(1 - \frac{R}{m}) - m - \sum_j c_j\}$. Substituting for w_i, $\Pi(m) = \frac{R}{2m}\{G^{-1}(1 - \frac{R}{m}) - (\sum_i c_i + m)\}$. The supply chain surplus can be written as $2\Pi(m) + R = \frac{R}{m}\{G^{-1}(1 - \frac{R}{m}) - \sum_i c_i\}$. Note that $G^{-1}(1 - \frac{R}{m}) = \sum_i w_i + m = p$, and $\frac{R}{m} = 1 - G(\sum_i w_i + m) = 1 - G(p) = $ Demand.

Hence, Supply chain surplus = Demand $\times (p - \sum_i c_i)$.

Thus the supply chain surplus is the same as in an integrated firm, and the supply chain is coordinated.

Chapter 3

Managing Capacity

3.1 Introduction

Supply chain capacity can be designed and organized in different ways
based on how it is built and the market environment. In most scenarios capacity can be understood in terms of the facility, equipment, and
workers. Restructuring capacity is one of the ways of responding to market driven changes. In the supply chain strategy matrix in Chapter 1
(Figure 1.2), observe that the supply chain processes of production and
capacity have high interactions with the market drivers such as, customized products, process yield, technology, quick response, and cost.
When Adidas (Seifert, 2003) wanted to customize shoes to individual
customers, they needed to alter the production process to respond to
variations in shoe-specifications from one customer to the next. Such
changes require restructuring of capacity, perhaps, from batch production to automation where cutting machines could be programmed to
switch to cutting different shapes. Customization also increases the
variability in the type and quantity of inbound material and, therefore,
may require more frequent deliveries. As a consequence the suppliers
would need to restructure their capacity to enable a more flexible production schedule. A rapid pace of technology change and improvements
in process yield, similarly, imply changes in production capacity — from

manual to automated and flexible. Clearly, the supply chain capacity must adapt to changes in demand, product portfolio, and customers' desire for quicker delivery. A firm may achieve responsiveness through updating the existing facilities, building new facilities, and sourcing from third parties (Waller, 2004).

Thus, the strategic fit between market drivers and the capability and positioning of the supply chain capacity becomes critical, as it determines the effectiveness of response to changes. The options available to the firm include capacity and demand allocation (Chapter 2), tailoring strategic initiatives such as made-to-stock and built-to-order to individual products, and investing in flexibility and in creating networked-capacities.

3.2   Types of Capacity

A production facility can be rigid, flexible, automated, lean, modular, and reconfigurable. It can operate as stand-alone or in a network. It can serve several markets from a central location or be dedicated to specific markets. It can specialize (focused) in few products or operate as a general purpose job shop. A portion of the capacity may be built in anticipation of demand (speculative capacity) while the remainder is postponed until after the demand is known (reactive capacity). Reactive capacity is usually reserved for products with high risk of mismatched supply and demand, permitting use of updated forecasts.

Manufacturing and service facilities can be fully or partially automated (capital intensive) or they can be run manually (labor intensive). There are two types of automation — hard and programmable. In hard automation, special-purpose equipment is used to perform tasks on high-volume products or services, involving repetitive operations. In such a facility it is hard to make adjustments to changes in processing sequence, or switch to different operations. In contrast, programmable automation permits the use of general purpose equipment and the sequence of processes can be easily altered. Programmable automation is thus very effective in producing small lots of customized products.

A flexible facility is designed to handle changes in work orders and tooling for several types of parts. It may be manual or it may use programmable automation. Examples of automated flexible systems are the programmable tool banks that support cutting machines, and programmable welding robots. A job shop, where workers are trained in multiple tasks, is an example of manual flexible facility. Flexibility permits rapid introduction of new products, and expansion to new market-regions and customers. Zara Inc (Ghemawat and Nueno, 2003) divides its products into fashion sensitive and time sensitive lines, and produces the fashion sensitive products in house using flexible technology. The time sensitive products are outsourced for large-volumes production at low cost. Li & Fung requires a great deal of flexibility in varying the size of its capacity by adding or deleting suppliers from its supply chain, based on customer orders. In a modular facility, tasks are grouped into modules with the objective of minimizing transfer of semi finished jobs between modules. Such modules are known as manufacturing cells. Groups of tasks based on their position upstream/downstream in the supply chain can also be considered modules. Examples are task-groups created for procurement, component manufacturing, and final assembly.

A focused facility strives for a narrow range of products, customers and processes. The result is a facility that is smaller, simpler and totally focused on few products or operations. However, focus need not imply absence of flexibility. An assembly line is a focused facility. A manufacturing cell, on the other hand, combines focus with flexibility.

The two extremes of facility-organization are market-specific facilities, each selling to its local market (decentralized); and a single facility supplying all markets (centralized). The generic configuration will clearly be the one that has facilities built in one subset of market locations, supplying to another subset of markets in addition to its own market, as shown in Figure 3.1. As manufacturing cost and demand of products may vary between locations, configuration of facilities and markets for different products need not be identical (Chakravarty, 2005a). Thus the links between facilities and markets (shown by arrows in Figure 3.1) would be different for different products, depending on whether the demand for a product in a certain market is satisfied with local production or with goods from other locations.

Market 1   Market 2   Market 3   Market 4

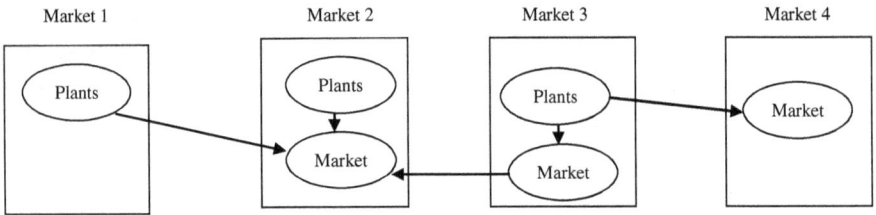

Fig. 3.1 Networked facilities.

3.3 Value of Capacity

The value of capacity is a concept useful in deciding whether the investment in a facility is justified. It is simply defined as the revenue one can earn from a given capacity of known size. However, as the revenue depends on several factors, some random and some controllable, its value may not be unique. Consider demand volatility. If the revenue that can be generated from the facility per unit of a product is $10, the value of capacity of size 200 would be $2000 only if the demand exceeds the capacity. However, if the probability of demand exceeding the total capacity is 0.4 and its average value when it does not exceed capacity is 150, the value of capacity with respect to demand would be $\{0.6(150) + 0.4(200)\}10 = 1700$. Note that the *marginal* value (of capacity) in demand is zero if demand exceeds capacity.

The manager can take certain actions to counter the demand uncertainty. He/she can configure the capacity to make it flexible for producing another product with demand that is negatively correlated with the first product. Then, together, these two products can ensure that the combined demand would be high even if the demand for one of the products dropped below the capacity of 200, and hence the value of capacity would increase. Obviously, the value would depend on demand variance, mean, and correlation and we would be interested in the sensitivity of value in demand parameters.

Next, we formalize the impact of demand volatility on the value of capacity. Consider demand ξ with distribution $F(\xi)$, variance σ^2, and mean μ. If the unit revenue is p, the marginal value of capacity with respect to demand would be p provided demand does not exceed the capacity; it is zero if demand exceeds capacity. Therefore, for a capacity of size Q, we express revenue as

$$\pi = p \int_{\xi \leq Q} \xi f(\xi) d\xi + pQ \int_{\xi \geq Q} f(\xi) d\xi, \quad E(\xi) = \mu, \quad V(\xi) = \sigma^2$$

To understand how the value may be related to demand volatility, we assume demand to be normally distributed, and rewrite revenue in terms of the standard normal variable z ($z = (\xi - \mu)/\sigma$) as

$$\pi = p \int_{z \leq \frac{Q-\mu}{\sigma}} (\mu + z\sigma)\varphi(z) dz + pQ \int_{z \geq \frac{Q-\mu}{\sigma}} \varphi(z) dz,$$

$$E(z) = 0, \quad \sigma(z) = 1$$

Note that $Q - \mu$ is the buffer capacity, $Q < \infty$, and $\sigma > 0$. Therefore, $\frac{Q-\mu}{\sigma} < \infty$. We may now study the marginal value of capacity in μ and σ.

$$\frac{\partial \pi}{\partial \mu} = p\Phi\left(\frac{Q-\mu}{\sigma}\right)$$

$$\frac{\partial \pi}{\partial \sigma} = p \int_{z \leq \frac{Q-\mu}{\sigma}} z\varphi(z) dz < pE(z) < 0$$

Note that $E(z) = 0$ by definition.

Thus, the marginal value of capacity *decreases* in demand volatility σ (VanMieghem, 2008) and it increases in μ. The marginal value may be thought of in terms of a "put" option that gives the option holder the right to sell capacity at an agreed upon price (option value). If increasing volatility decreases the marginal value of capacity sufficiently below the option value, the owner of the put option will consider selling the capacity.

Next, note that the rate of increase in revenue in capacity Q, $\frac{\partial \pi}{\partial Q} = p\{1 - \Phi(\frac{Q-\mu}{\sigma})\} \geq 0$ and $\frac{\partial^2 \pi}{\partial Q^2} = -\frac{p}{\sigma}\phi(\frac{Q-\mu}{\sigma}) \leq 0$. Note that π increases, at a decreasing rate, in Q. Denoting the marginal cost of capacity to be c, the optimal capacity is expressed as $p\{1 - \Phi(\frac{Q^*-\mu}{\sigma})\} = c$, Thus,

$$Q^* = \sigma\Phi^{-1}\left(1 - \frac{c}{p}\right) + \mu \quad \text{and} \quad \frac{\partial Q^*}{\partial \sigma} = \Phi^{-1}(1 - c/p)$$

It follows that the optimal size of buffer capacity $Q^* - \mu$ increases in σ. However, as the option value of capacity decreases in σ, it would not be advisable to increase the buffer capacity beyond a threshold value. At that point the firm would need to consider adding capacity that can be easily rolled back such as overtime-capacity, additional shift, and outsourcing.

3.4   Capacity Allocation

We have discussed capacity allocation for the airline industry in Chapter 2. Recollect that the major decision in that analysis was whether to sell seats to customers who were not willing to pay a high fare but were certain to buy, or to hold out for high paying but uncertain customers. It did not involve restructuring of capacity or innovative pricing.

For consumer products, the scenario is a little different, as the product is taken through multiple processing steps before completion and it requires production capacity at each step. In addition, customers may choose different products that may require different processes. The product a customer wants may be in stock, in which case he/she would get instant delivery; or it may be out of stock, in which case he/she would have to wait for it to be built in real time. Correspondingly, there are two clear choices for the supplier: (i) make-to-stock (MTS) where inventories of each product type are built based on estimates of demand, or (ii) build-to-order (BTO) where the products, if sold out, are built in real time based on actual demands (Arreola-Risa and DeCroix, 1998b). Rajagopalan (2002) discusses which products should be made to stock, and which built to order. To minimize customer wait time in BTO, the supplier may consider a third option where the product goes through a subset of common processes before demands materialize, and customized steps (for the ith product) are completed based on actual demand. We call this make to stock and order (MTSO), which is a way of decoupling the upstream production processes from the variability in customer demand downstream (Stratton and Warburton, 2003).

While MTS ensures demand fulfillment for customers who arrive early, MTSO facilitates fulfillment of late orders. As the demand is not known in advance, a policy that combines MTS with MTSO can be optimal. To implement this, capacity is allocated to two buckets: bucket 1 with finished products caters to the demand of early customers, and bucket 2 with the generic product is earmarked for the late customers.

Consider a scenario with n final products and demand ξ_i, $i = 1, n$. The supplier builds MTS inventories of Q_i at unit cost c_i, and holds them in bucket 1. The supplier builds and holds Q units of the "generic" product (unit cost c) in stock in bucket 2. Of this Q units, q_i units are reserved for conversion to product $i (\sum_{i=1,n} q_i \leq Q)$, at cost r_i per unit, $c \leq c_i \leq c + r_i$. If ξ_i lies in the range $Q_i \leq \xi_i \leq Q_i + q_i$, $Q - \sum_{i=1,n} \min(\xi_i - Q_i, q_i)$ units of the generic product is left over in inventory. Demand in excess of $Q_i + q_i$ is lost. The supplier needs to determine the optimal mix of MTS capacity Q_i and MTSO capacity q_i.

Assuming zero residual value of inventory at the end of the planning horizon, and letting p_i to be the unit sales price, the supplier's expected profit is written as

$$\pi = \begin{vmatrix} \sum_i p_i \left\{ \int_{\xi_i \leq Q_i + q_i} \xi_i f_i(\xi_i) d\xi_i + (Q_i + q_i) \int_{\xi_i \geq Q_i + q_i} f_i(\xi_i) d\xi_i \right\} \\ -\sum_i r_i \left\{ \int_{Q_i \leq \xi_i \leq Q_i + q_i} (\xi_i - Q_i) f_i(\xi_i) d\xi_i \right. \\ \left. + q_i \int_{\xi_i \geq Q_i + q_i} f_i(\xi_i) d\xi_i \right\} - \sum_i c_i Q_i - cQ \end{vmatrix}$$

Subject to

$$\sum_i q_i \leq Q, \quad c \leq c_i \leq c + r_i \leq p_i$$

Using Lagrange multiplier λ, the first order conditions in q_i and Q_i are written as

$$(p_i - r_i)\{1 - F_i(Q_i + q_i)\} - \lambda = 0$$
$$-(p_i - r_i)F_i(Q_i + q_i) - r_i F_i(Q_i) + p_i - c_i = 0$$

Solving the above equations we have,

$$Q_i^* = F_i^{-1}\left(1 - \frac{c_i - \lambda}{r_i}\right), \quad \text{and} \quad q_i^* = F_i^{-1}\left(1 - \frac{\lambda}{p_i - r_i}\right) - Q_i$$

Note that Q_i^* increases and q_i^* decreases in λ, and λ decreases in Q. This is intuitive, as a decrease in Q causes reduction in q_i^*, and Q_i^* increases to make up for the reduction in q_i^*. As $q_i \neq \infty$, it must be the case that $\lambda > 0$, implying $\sum_i q_i = Q$. We thus solve for λ using,

$$\sum_i \left\{ F_i^{-1}\left(1 - \frac{\lambda}{p_i - r_i}\right) - F_i^{-1}\left(1 - \frac{c_i - \lambda}{r_i}\right) \right\} = Q,$$

$$\text{and} \quad \lambda < \frac{(p_i - r_i)c_i}{p_i} \quad (\text{as } Q > 0).$$

Using the above value of λ we can solve for Q_i^* and q_i^* in terms of Q. Note that as λ is defined for a specific Q, it is a dependent variable. From the first order condition $\lambda(Q^*) = c$, we obtain optimal Q by noting that $\lambda(Q^*)$ must satisfy

$$\sum_i \left\{ F_i^{-1}\left(1 - \frac{\lambda(Q^*)}{p_i - r_i}\right) - F_i^{-1}\left(1 - \frac{c_i - \lambda(Q^*)}{r_i}\right) \right\} = Q^*.$$

Thus,

$$Q^* = \sum_i \left\{ F_i^{-1}\left(1 - \frac{c}{p_i - r_i}\right) - F_i^{-1}\left(1 - \frac{c_i - c}{r_i}\right) \right\}$$

It follows that $F_i^{-1}(1 - c/(p_i - r_i))$ represents the maximum amount of product i that can be bought $(= Q_i^* + q_i^*)$. Note that q_i is a target value and is not an actual production capacity. Therefore, letting $\sum_i q_i$ to exceed Q is akin to the practice of overbooking in the airline industry. It may, of course, result in unfilled orders due to the shortage of generic products. The company can offer rain-checks or provide incentives to customers to wait longer for all processing steps to be completed.

3.4.1   Waiting Time for MTSO

Denote the time a customer is willing to wait as t, and $\text{Prob}(t \leq T) = G(T)$ where T is the actual wait time, so that utility can be defined as

$u(T) = 1 - G(T)$. The profit would be

$$
\pi = \left| \begin{array}{l} \sum_i p_i \left\{ \int_{\xi_i \leq Q_i} \xi_i f_i(\xi_i) d\xi_i + Q_i \int_{\xi_i \geq Q_i} f_i(\xi_i) d\xi_i \right\} - \sum_i c_i Q_i - cQ \\[2ex] + u(T) \sum_i (p_i - r_i) \left\{ \int_{Q_i \leq \xi_i \leq Q_i + q_i} (\xi_i - Q_i) f_i(\xi_i) d\xi_i \right. \\[2ex] \left. + q_i \int_{\xi_i \geq Q_i + q_i} f_i(\xi_i) d\xi_i \right\} \end{array} \right|
$$

It can be shown that

$$
q_i = F_i^{-1} \left(1 - \frac{\lambda}{u(T)(p_i - r_i)} \right) - Q_i,
$$

$$
Q_i = F_i^{-1} \left(1 - \frac{c_i - \lambda}{p_i - u(T)(p_i - r_i)} \right), \quad \text{and}
$$

$$
Q = \sum_i F_i^{-1} \left(1 - \frac{\lambda}{u(T)(p_i - r_i)} \right)
$$

$$
- \sum_i F_i^{-1} \left(1 - \frac{c_i - \lambda}{p_i - u(T)(p_i - r_i)} \right)
$$

As $u(T)$ decreases in T, Q_i increases and Q decreases in T. An increase in T implies longer wait in MTSO fulfillment and the company counters it by increasing the size of the MTS inventory Q_i.

3.4.2 Proactive and Reactive Capacity

It is well known that it costs far less (low unit cost) to build proactive-capacity before the demand is known. However, the cost of redundancy in capacity, built as a hedge against upside risk, can be high. Reactive capacity, on the other hand, is built in real time after demand is known (ex post). Although resource utilization is high, the cost of procuring resources at short notice can be very high. Cattani et al. (2008) following the work of Eynan and Rosenblatt (1995) establish conditions, involving capacity costs, for investment in both reactive and proactive capacities (which is also known as speculative capacity). Fisher and Raman (1996) discuss a model where an initial capacity is built based on a forecast prior to realization of demand, with a provision of adding capacity later when demand became known. Clearly a reactive capacity is more suited to high demand volatility, but it needs to function with proactive capacity to keep costs down.

There are other competitive advantages in proactive and reactive capacities that we have not included. As capacity decisions are made after demand is observed in a reactive mode, companies can benefit by exploiting the option-value of waiting. Zara Inc invests in capacity in two stages. In stage 1, it builds a small capacity for a new product and monitors how customers respond to this product. It uses this information to assess demand for the product and build (or discontinue the product) a larger facility in stage 2, appropriate for the assessed demand. This can be interpreted as Zara creating a new market for its new product by leveraging its stage 1 facility. It neither creates extensive databases of past demands nor uses sophisticated software for analyzing demand patterns. There is considerable risk to Zara if competitors get ahead of it with a similar product and build a large facility in stage 1. Zara stands to lose customers. Therefore Zara's strategy comprises a second feature: rapid development of new products. If a product fails in stage 1, it can introduce the next product quickly with another "stage 1 investment".

Other elements of the option-value of waiting are the opportunity to use emerging new technology, and the opportunity to learn and make more effective use of distribution channels and customer relationship.

Perhaps the biggest advantage of proactive capacity is the ability to mitigate risk from demand volatility. In addition, it can exploit the first-mover advantage by capturing a large market share. The company can also use proactive capacity as an entry barrier against new competitors.

3.5   Flexible Capacity

Flexible capacity is perhaps the most effective option in providing customization in a market driven environment (Boute et al., 2009). It enables firms to switch production from one product to another with little delay and negligible setup cost. Therefore, a shortfall in demand of one product can be made up by other products, all using the same facility.

3.5.1   Capacity Restructuring for Flexibility

The question most companies are interested in is how to assess the value of flexibility, and how to incorporate flexibility in manufacturing or service. Consider flexible capacity of size R that can produce two products, with demands ξ_i, distribution $F(\xi_i)$, and unit price $p_i, i = 1, 2$. When the combined demand $\sum_{i=1,2} \xi_i$ exceeds R, an allocation rule is used to determine capacity for the competing products. We assume that p is the revenue *per unit capacity*, based on a given allocation rule. Clearly, p will be a convex combination of $p_i, i = 1, 2$. The firm's profit is written as

$$
\pi = \int_{\xi_2 \leq R} \int_{\xi_1 \leq R - \xi_2} (p_1 \xi_1 + p_2 \xi_2) f_1(\xi_1) f_2(\xi_2)
$$

$$
+ pR \int_{\xi_2} \int_{\xi_1 \geq R - \xi_2} f_1(\xi_1) f_2(\xi_2) - cR
$$

$$
\frac{\partial \pi}{\partial R} = p \left[\int_{\xi_2} \{1 - F_1(R - \xi_2) - R f_1(R - \xi_2)\} f_2(\xi_2) d\xi_2 \right] - c
$$

Assuming $F(\cdot)$ to be IGFR, $1 - F_1(R - \xi_2) - R f_1(R - \xi_2)$ can be shown to decrease from a positive to a negative value in R for any value of ξ_2, implying π is unimodal in R. Thus, from the first order condition R^* (optimal R) is determined from,

$$
\int_{\xi_2} \{1 - F_1(R^* - \xi_2) - R^* f_1(R^* - \xi_2)\} f_2(\xi_2) d\xi_2 = \frac{c}{p}
$$

As p is a function of the allocation rule, R^* will be different for different allocation rules. Clearly, the optimal allocation rule would be to first assign capacity to product i and then to j such that $\frac{p_i}{a_i} \geq \frac{p_j}{a_j}$, where p_i is the unit revenue and a_i the unit capacity needed for product i. Using this allocation rule, we may express p as $p(R) = \int_{\xi_i \leq R} \{ \int_{\xi_j \leq R - \xi_i} (p_i \xi_i + p_j \xi_j) f_j(\xi_j) d\xi_j \} f_i(\xi_i) d\xi_i$. Thus the first order condition for R^* above must be solved iteratively for R and $p(R)$.

To assess the value of flexible capacity we explore an aggregate analysis by substituting the two products with an equivalent product such that $\xi = \xi_1 + \xi_2$, $E(\xi) = \mu = \mu_1 + \mu_2$ and

$V(\xi) = \sigma^2 = \sigma_1^2 + \sigma_2^2 + 2\text{Cov}(\xi_1, \xi_2)$. The profit expression would be,

$$\pi = p_e \left\{ \int_{\xi \leq R} \xi f(\xi) + R \int_{\xi \geq R} f(\xi) \right\} - cR$$

where p_e is the unit revenue of the equivalent product.

Using news-vendor analysis, the optimal value of capacity size would satisfy

$$F(R) = 1 - \frac{c}{p_e}$$

$F(R)$ is transformed to,

$$\Phi \left(\frac{R - (\mu_1 + \mu_2)}{\sqrt{\sigma_1^2 + \sigma_2^2 + 2\text{Cov}(\xi_1, \xi_2)}} \right) = 1 - \frac{c}{p_e}$$

Flexible capacity provides risk pooling advantage, as $\sqrt{\sigma_1^2 + \sigma_2^2 + 2\text{Cov}(\xi_1, \xi_2)} \leq \sigma_1 + \sigma_2$. Thus, we may assess the value of flexibility as $\sigma_1 + \sigma_2 - \sqrt{\sigma_1^2 + \sigma_2^2 + 2\text{Cov}(\xi_1, \xi_2)}$. Note that if ξ_1 is independent of ξ_2, $\text{Cov}(\xi_1, \xi_2) = 0$, and the value of flexibility will equal $\sigma_1 + \sigma_2 - \sqrt{\sigma_1^2 + \sigma_2^2}$. It can be verified that the value of flexibility decreases in positive demand correlation. Therefore, if the products are perfectly correlated positively so that the covariance equals one, the value of flexibility reduces to zero. Similarly, if they are perfectly correlated negatively, the flexibility value equals $2\sigma_1$ or $2\sigma_2$.

To explore the impact of demand volatility, we note $\frac{\partial R}{\partial \sigma} = \frac{R - (\mu_1 + \mu_2)}{\sigma}$. Comparing with the scenario of two dedicated facilities,

$$\frac{\partial R}{\partial \sigma} - \left(\frac{\partial R_1}{\partial \sigma_1} + \frac{\partial R_2}{\partial \sigma_2} \right) = \frac{R - (\mu_1 + \mu_2)}{\sigma}$$

$$- \left(\frac{R_1 - \mu_1}{\sigma_1} + \frac{R_2 - \mu_2}{\sigma_2} \right) \leq 0$$

Thus we clearly establish that the impact of demand volatility on capacity size can be decreased by using a flexible capacity.

In the service sector, flexibility is incorporated in the facility by enabling it to provide a portfolio of services. For example, courier services such as FedEx can provide multiple services in parcel delivery by size, weight, and delivery time. If more customers opt for overnight delivery, FedEx can expand the capacity of its aircraft fleet, or lease additional cargo space on routinely scheduled airline flights.

3.5.2 Health Care Application

In a hospital, the specialist is like a dedicated facility and the general practitioner (GP) acts as a flexible facility. However, there are important differences. First, unlike manufacturing, job-flow is reversed; it flows from a flexible facility to a dedicated facility. Patients first consult with the GP (flexible) and then with a specialist (dedicated), if needed. The GP, because of lack of specialization, can function only within a specific range of medical knowledge. If the demand (patient's disease) crosses the GP's knowledge threshold, the patient is referred to the specialist. Second, unlike manufacturing where the capacity of first stage in the "flow" is exhausted before the second stage (stage with flexibility) is invoked, in certain scenarios the GP can be idle while the specialist is busy. Therefore, the demand for hospital services is defined by two attributes: the number of patients ξ_i (distribution $F_i(\xi_i)$) in patient category i who need hospital service per period, and e_i the level of physician-expertise needed to treat the patients in category i with distribution $H_i(e_i)$. Let E_i represent the GP's expertise threshold for patients in category i, so that the probability that a patient is referred to the specialist is $1 - H_i(E_i)$. Let the planned capacities of GP and specialist in category i be R_G and R_{Si}, respectively.

We assume that if the total patient demand for GP-service exceeds R_G, patients in category i are allocated capacity $\alpha_i R_G$, $\sum_i \alpha_i = 1$. Therefore the profit of a hospital that serves only two categories of patients would be

$$
\pi = \left| \begin{aligned}
&p_G \int_{\xi_2 \leq \frac{R_G}{H_2(E_2)}} \int_{\xi_1 \leq \frac{R_G - \xi_2 H_2(E_2)}{H_1(E_1)}} \{\xi_1 H_1(E_1) \\
&\quad + \xi_2 H_2(E_2)\} f_1(\xi_1) f_2(\xi_2) d\xi_1 d\xi_2 \\
&+ p_G R_G \sum_{i=1,2} \alpha_i \int_{\xi_i H_i(E_i) \geq \alpha_i R_G} f_i(\xi_i) d\xi_i - c_G R_G - \sum_{i=1,2} c_{Si} R_{Si} \\
&+ \sum_{i=1,2} p_{Si} \left\{ \int_{\xi_i\{1-H_i(E_i)\} \leq R_{Si}+\alpha_i R_G} [\xi_i\{1 - H_i(E_i)\} \\
&\quad - \alpha_i R_G] f_i(\xi_i) d\xi_i + R_{Si} \int_{\xi_i\{1-H_i(E_i)\} \geq R_{Si}} f_i(\xi_i) d\xi_i \right\}
\end{aligned} \right.
$$

The capacity related decision variables are R_G and R_{Si}, $\forall i$. It would be interesting to study E_i as a decision variable as well — it would be useful in establishing guidelines as to when a patient should be referred to a specialist. It would also have implications for capacity planning and training programs for the GP.

3.5.3   Requisite Flexibility

How much flexibility is adequate? Consider three-product, three-plant production system. To cope with demand volatility, flexible facilities are being considered. A facility could be dedicated to one product (no flexibility) or, at the other extreme, it could be configured to produce all three products (full flexibility), as shown in Figure 3.2a.

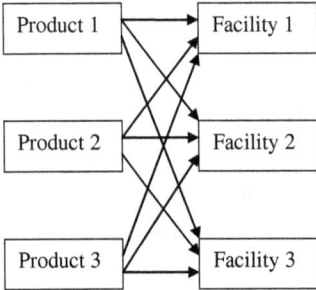

Fig. 3.2a Full flexibility.   Fig. 3.2b Partial flexibility.

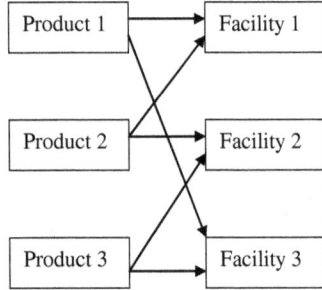

Jordan and Graves (1995) showed that partial flexibility, shown in Figure 3.2b, where each plant produces only 2 products could be adequate for satisfying production needs. Note in Figure 3.2b, if the demand for product 2 increases,capacities at plants 1 and 2 may become saturated, resulting in lost sales even if facility 3 has excess capacity. Jordan and Graves establish that while capacity utilization with full flexibility increases in relation to that with partial flexibility, the increase is only marginal. The corresponding increase in cost of building the full-flexibility facilities can be significantly higher.

3.6   Networked Capacity

A capacity network enables coordination of capacity related decisions at different stages of a supply chain. It implies collaboration and not a decision hierarchy. It helps moderate the impact of demand volatility by offsetting capacity shortage at one location with capacity surplus at another. Such networks may exist in relationships that are vertical, horizontal, or "regional". A vertical network defines the relationship in a vertical chain, such as that between a retailer and a supplier (Wal-Mart and P&G), or an assembler and a component manufacturer for logistics planning (Ford and Johnson Controls). In a horizontal network competitors such as airlines collaborate through code-sharing and alliances. Regional networks emerge to leverage cost and/or purchasing-power differentials between different regions of the world.

3.6.1   Vertical Network

Consider a vertical network between a manufacturer and a supplier. Assume that the manufacturer can reduce its unit cost by incorporating new technology and would like to exploit it by increasing production volume. However, if the supplier's cost structure remains unchanged, he would continue to supply components at the same rate, thus causing a mismatch. To delve deeper, consider the newsvendor solution, where the manufacturer determines her production quantity Q_1 to be $F(Q_1) = 1 - \frac{c_m}{p}$, and the supplier (using the same demand distribution) determines his production quantity to be $F(Q_2) = 1 - \frac{c_s}{p}$, where c_m is the unit manufacturing cost and c_s is the supplier's unit cost. There will be a mismatch as $Q_1 \neq Q_2$, if $c_m \neq c_s$. In a coordinated solution, the synchronized quantity is determined as $F(Q) = 1 - \frac{c_m + c_s}{p}$, which is smaller than both Q_1 and Q_2. Therefore, $\frac{\partial Q}{\partial \sigma} = \frac{Q-\mu}{\sigma} \leq \frac{\partial Q_i}{\partial \sigma} = \frac{Q_i-\mu}{\sigma}, i = 1, 2$. That is, *the impact of demand volatility is lower in the coordinated decision than in the case of individual decisions.*

In the decentralized-decision scenario, the supplier gets to decide his wholesale price. However, if decisions are made sequentially (Stackelberg), it can be shown that the supplier can increase the wholesale price whenever the manufacturer achieves a productivity driven reduction in manufacturing cost. Thus, the supplier benefits from the free-rider effect, as increase in the wholesale price w is not contingent upon improvement in the supplier's productivity.

3.6.2   Lateral Network

A lateral (horizontal) network can serve as a substitute for flexible facility by providing for transshipment of finished products between two networked facilities. The emergence of competing carriers with fast deliveries has made it possible to efficiently move items between facilities. Rudi et al. (2001) show that the firms can increase their profits when transshipment is allowed. Grahovac and Chakravarty (2001) consider a supply chain with a distribution center (DC) supplying expensive parts (such as aircraft components) to n retailers. They establish that by permitting transshipment among retailers, the inventory held at the DC decreases while the inventory held by an individual retailer increases. This is another example of free-rider, as the decrease in DC's inventory is not productivity induced.

If the firms do not sell identical products, transshipment would not be very useful for creating flexibility. Exchange of capacity between two networked facilities (firm 1's product can be manufactured in firm 2's facility and vice versa) would be the way of providing flexibility. As an example, Renault's plant in Brazil is set to produce two additional Nissan vehicles, while Nissan's plant in South Africa builds two additional Renault vehicles. In 2010, a total of 11 vehicles will be cross-manufactured (Ghosn, 2009). Capacity sharing in network services is implemented through capacity swaps and redundancy agreements (Yankee, 2001). Another example is the telecommunication tower, which is shared by multiple network operators (Chakravarty, 2009b).

As in Chakravarty and Zhang (2007), consider a company with capacity R; the company can buy or sell capacity in the market at price p. Assume the capacity, after exchange has taken place, to be Q. The company's profit is written as

$$\pi(Q) = -(p+t)\min(Q-R,0) + p\min(R-Q,0)$$
$$+ m\left\{\int_{\xi \leq Q} \xi f(\xi)d\xi + Q \int_{\xi \geq Q} f(\xi)d\xi\right\}$$

where m is the revenue earned per unit capacity, and t is the cost of converting the purchased capacity to the buyer's specifications.

Note that

$$\frac{\partial \pi}{\partial Q} = \begin{vmatrix} -p + m\{1 - F(Q)\}, & \text{if } Q < R \\ -(p + t) + m\{1 - F(Q)\}, & \text{if } Q > R \end{vmatrix}$$

We can solve for Q from the corresponding first order condition. Thus,

$$Q = \begin{vmatrix} F^{-1}\left(1 - \frac{p}{m}\right), & \text{if } m\bar{F}(Q) < p \\ R, & \text{if } p \le m\bar{F}(Q) \le p + t \\ F^{-1}\left(1 - \frac{p+t}{m}\right), & \text{if } m\bar{F}(Q) > p + t \end{vmatrix}$$

Clearly the company will sell $R - F^{-1}(1 - \frac{p}{m})$ units of capacity if $p > m\bar{F}(Q) = p_s$, and buy $F^{-1}(1 - \frac{p+t}{m})$ units if $p < m\bar{F}(Q) - t = p_b$. It will do nothing if $m\bar{F}(Q) - t \le p \le m\bar{F}(Q)$. That is, if the market price drops below p_b the firm will purchase capacity, and if it exceeds p_s it will sell capacity.

Capacity Trading

We consider trading of capacity between two firms. If firm 1 sets price p such that $p_{1s} \le p \le p_{2b}$, firm 2 will purchase a quantity $q = F^{-1}(1 - \frac{p+t_{12}}{m})$, t_{12} indicates capacity conversion from firm 1 to 2. As $q = F^{-1}(1 - \frac{p+t_{12}}{m})$ implies $p = m\{1 - F(q)\} - t_{12}$, firm 1 (seller) will maximize

$$\pi_s(q) = q\{m_2\{1 - F(R_2 + q)\} - t_{12}\}$$
$$+ m_1 \left\{ \int_{\xi \le R_1 - q} \xi f(\xi) d\xi + (R_1 - q) \int_{\xi \ge R_1 - q} f(\xi) d\xi \right\}$$
$$p_{1s} \le m_2\{1 - F(R_2 + q)\} - t_{12} \le p_{2b} \quad \text{and} \quad q \le R_1$$

It can be shown (Chakravarty and Zhang, 2007) that q will be the solution of the following equation,

$$1 - F(R_2 + q) - qf(R_2 + q) = \frac{m_1}{m_2}\{1 - F(R_1 - q)\} + \frac{t_{12}}{m_2}$$

Thus, the amount of capacity bought in a lateral network can be determined uniquely.

3.7   Outsourced Capacity

Most companies outsource their non critical tasks, primarily, to reduce production costs. The outsourcing partner, by aggregating orders from multiple customers, can achieve a lower unit cost due to scale-economy. In a market driven environment an equally important factor to consider would be whether outsourcing increases a firm's vulnerability to demand volatility.

There are two approaches in most outsourcing decisions: outsource capacity and outsource capability. Capacity outsourcing can be implemented in two different ways: outsource the stable part of demand, or outsource the volatile part of demand. In outsourcing volatile demand component, demand exceeding Q is outsourced. The converse is the case in outsourcing the stable demand component where up to first Q units of demand are outsourced, as in the case of Griffin Manufacturing (Chapter 2) where stable demand component is outsourced to the manufacturer in Honduras. Note that outsourcing the volatile demand component is very common.

Capability outsourcing is used by a company that specializes in certain tasks, and outsources tasks it does not do itself. For example, a company may outsource component manufacturing. It decreases component cost but may increase the risk of components not being delivered on time, delaying all subsequent tasks such as assembly. Most offshore outsourcing is of this type. System integrators, a new job description in outsourcing companies, specialize in coordinating different parts of a job before and after they are completed by different partners worldwide. Zara uses capability outsourcing by keeping fashion sensitive products in house and outsourcing price sensitive products globally.

3.7.1   Outsourcing the Volatile Demand Component

Consider an outsourcing partner possessing capacity R. The company outsources all demand exceeding capacity Q_R. Assuming partner's wholesale price w, unit cost of inhouse capacity c, unit sales price p, and a random demand ξ, the company's profit is expressed as

$$\pi_1 = pE_\xi\{\min(\xi, Q_R + R)\} - wE_{\xi \geq Q_R}\{\min(\xi - Q_R, R)\} - cQ_R$$

As in Chapter 2, we obtain optimal Q_R from the first order condition as,

$$(p - w)F(Q_R^* + R) + wF(Q_R^*) = p - c, \tag{3.1}$$

Equation (3.1) can be rewritten, in terms of a standardized Normal distribution, as

$$(p - w)\Phi\left(\frac{Q_R^* + R - \mu}{\sigma}\right) + w\Phi\left(\frac{Q_R^* - \mu}{\sigma}\right) = p - c, \tag{3.2}$$

Denoting Q^* as the optimal capacity without outsourcing, it follows from appendices I and II, respectively,

$$\frac{\partial Q_R^*}{\partial \sigma} \geq \frac{\partial Q^*}{\partial \sigma}, \quad \text{and} \quad Q_R^* \leq Q^* \leq Q_R^* + R, \quad \text{for } R > 0$$

As $Q_R^* + R \geq Q^*$, we conclude that (a) outsourcing increases the total capacity needed in the system, and (b) it increases volatility of the in-house capacity.

It can be verified that the maximum profit of the buyer for known R and w, would be

$$\pi_1(R) = (p - c)Q_R^* + (p - w)R - p \int_{\xi \leq Q_R^* + R} F(\xi)$$

$$+ w \int_{Q_R^* \leq \xi \leq Q_R^* + R} F(\xi)$$

Clearly, outsourcing will be profitable only if $\frac{\partial \pi_1(R)}{\partial R} > 0$ at $R = 0$. Assuming that is the case, the company will need to resolve the tradeoff between increased profit and increased volatility in capacity to decide whether to outsource and, if so, how much.

3.7.2 Outsourcing Stable Demand

A company may outsource the stable part of the demand to a low cost producer, and produce the volatile demand inhouse. This is a variation of the Griffin Manufacturing case in Chapter 2. Thus, the in-house facility is used only if the outsourcing partner's capacity R is exhausted. The company's profit is expressed as,

$$\pi_2 = pE_\xi\{\min(\xi, Q_R + R)\} - wE_\xi\{\min(\xi, R)\} - c_2 Q_R$$

From the first order condition,

$$F(Q_R^* + R) = 1 - \frac{c_2}{p}$$

Using Normal distribution, we express it as $\Phi(\frac{Q_R^* + R - \mu}{\sigma}) = 1 - \frac{c_2}{p}$, so that

$$Q_R^* + R = \sigma \Phi^{-1}\left(1 - \frac{c_2}{p}\right) + \mu = Q^* \quad \text{and}$$

$$\frac{\partial Q_R^*}{\partial \sigma} = \Phi^{-1}\left(1 - \frac{c_2}{p}\right) = \frac{\partial Q^*}{\partial \sigma}$$

Therefore, the total system capacity as well as the exposure to demand volatility remains unchanged in R.

The maximum profit of the buyer for given R and w would be

$$\pi_2(R) = (p - c_2)Q_{R2}^* + (p - w)R - p \int_{\xi \leq Q_{R2}^* + R} F(\xi) + w \int_{\xi \leq R} F(\xi)$$

The company can decide which form of outsourcing to use, volatile or stable component, based on the value of $\pi_1(R) - \pi_2(R)$. Note that

$$\pi_1(R) - \pi_2(R) = \begin{vmatrix} (p - c)Q_R^* - (p - c_2)Q_{R2}^* \\ + p\left\{ \int_{\xi \leq Q_{R2}^* + R} F(\xi) - \int_{\xi \leq Q_R^* + R} F(\xi) \right\} \\ + w\left\{ \int_{Q_R^* \leq \xi \leq Q_R^* + R} F(\xi) - \int_{\xi \leq R} F(\xi) \right\} \end{vmatrix}$$

3.7.3   Overtime Production

Consider the scenario where the firm uses overtime instead of outsourcing. Assume that the firm builds a facility of capacity Q at a unit cost c, and pays workers a wage of w per unit capacity. Assume the firm can use a fraction α of the capacity in overtime, and the overtime wage is

$w_o(> w)$. The firm's profit is written as

$$
\pi = \left| \begin{array}{l} p\left\{ \displaystyle\int_{\xi \leq Q+\alpha Q} \xi f(\xi) + (Q+\alpha Q) \int_{\xi \geq Q+\alpha Q} f(\xi) \right\} \\[2ex] \quad -w_o \left\{ \displaystyle\int_{Q \leq \xi \leq Q+\alpha Q} (\xi - Q)f(\xi) \right. \\[2ex] \quad \left. +\alpha Q \displaystyle\int_{\xi \geq Q+\alpha Q} f(\xi) \right\} - cQ - wQ \end{array} \right.
$$

From the first order condition in Q,

$$
(1+\alpha)(p-w_o)F((1+\alpha)Q^*) + w_o F(Q^*) = \alpha(p-w_o) + p - c - w
$$

Using standardized Normal distribution, we have

$$
(1+\alpha)(p-w_o)\Phi\left(\frac{(1+\alpha)Q^* - \mu}{\sigma}\right) + w_o\Phi\left(\frac{Q^* - \mu}{\sigma}\right)
$$

$$
= \alpha(p-w_o) + p - c - w
$$

As in the case of outsourcing, we can show that

$$
\frac{\partial Q^*_{(\alpha=0)}}{\partial \sigma} \leq \frac{\partial Q^*_{(\alpha>0)}}{\partial \sigma}
$$

3.8   Appendix I

From (3.2),

$$
\frac{\partial Q^*_R}{\partial \sigma} = \frac{\frac{w}{\sigma}\left(\frac{Q^*_R-\mu}{\sigma}\right)\phi\left(\frac{Q^*_R-\mu}{\sigma}\right) + \frac{p-w}{\sigma}\left(\frac{Q^*_R+R-\mu}{\sigma}\right)\phi\left(\frac{Q^*_R+R-\mu}{\sigma}\right)}{\frac{w}{\sigma}\phi\left(\frac{Q^*_R-\mu}{\sigma}\right) + \frac{p-w}{\sigma}\phi\left(\frac{Q^*_R+R-\mu}{\sigma}\right)}
$$

It can be easily shown that

$$
\frac{Q^*_R - \mu}{\sigma} \leq \frac{\frac{w}{\sigma}\left(\frac{Q^*_R-\mu}{\sigma}\right)\phi\left(\frac{Q^*_R-\mu}{\sigma}\right) + \frac{p-w}{\sigma}\left(\frac{Q^*_R+R-\mu}{\sigma}\right)\phi\left(\frac{Q^*_R+R-\mu}{\sigma}\right)}{\frac{w}{\sigma}\phi\left(\frac{Q^*_R-\mu}{\sigma}\right) + \frac{p-w}{\sigma}\phi\left(\frac{Q^*_R+R-\mu}{\sigma}\right)}
$$

$$
\leq \frac{Q^*_R + R - \mu}{\sigma}
$$

That is,

$$
\frac{Q^*_R - \mu}{\sigma} \leq \frac{\partial Q^*_R}{\partial \sigma} \leq \frac{Q^*_R + R - \mu}{\sigma}
$$

Note that when $R = 0$ (implying no outsourcing), $Q_R^* = Q^*$, where $Q^* = \sigma \Phi^{-1}(1 - \frac{c}{p}) + \mu$, and $\frac{\partial Q^*}{\partial \sigma} = \frac{Q^* - \mu}{\sigma}$. Therefore, $\frac{Q_{R(R=0)}^* - \mu}{\sigma} = \frac{\partial Q^*}{\partial \sigma}$. Hence,

$$\frac{Q_{R(R=0)}^* - \mu}{\sigma} = \frac{\partial Q^*}{\partial \sigma} \le \frac{\partial Q_R^*}{\partial \sigma} \le \frac{Q_R^* + R - \mu}{\sigma}$$

Thus with $R > 0$, we have

$$\frac{\partial Q^*}{\partial \sigma} \le \frac{\partial Q_R^*}{\partial \sigma} \le \frac{Q_R^* + R - \mu}{\sigma}$$

3.9   Appendix II

From (3.2),

$$\Phi\left(\frac{Q_R^* + R - \mu}{\sigma}\right)$$

$$= 1 - \frac{c}{p} + \frac{w}{p}\left\{\Phi\left(\frac{Q_R^* + R - \mu}{\sigma}\right) - \Phi\left(\frac{Q_R^* - \mu}{\sigma}\right)\right\}$$

$$\ge 1 - \frac{c}{p} = \Phi\left(\frac{Q^* - \mu}{\sigma}\right), \qquad\qquad \text{II.1}$$

Therefore,

$$Q^* = \sigma \Phi^{-1}\left(1 - \frac{c}{p}\right) + \mu \le Q_R^* + R$$

Note that (II.1) can be rewritten as,

$$\Phi\left(\frac{Q_R^* - \mu}{\sigma}\right) = \frac{p - c}{w} - \frac{p - w}{w}\Phi\left(\frac{Q_R^* + R - \mu}{\sigma}\right)$$

$$= \frac{p - c}{w} - \frac{p - w}{w}\Phi\left(\frac{Q_R^* - \mu}{\sigma}\right) - x$$

where

$$x = \frac{p - w}{w}\left\{\Phi\left(\frac{Q_R^* + R - \mu}{\sigma}\right) - \Phi\left(\frac{Q_R^* - \mu}{\sigma}\right)\right\}$$

Therefore, $\Phi(\frac{Q_R^* - \mu}{\sigma}) = 1 - \frac{c}{p} - x \le 1 - \frac{c}{p}$, implying $Q_R^* \le \sigma \Phi^{-1}(1 - \frac{c}{p}) + \mu = Q^*$.

Chapter 4

Product and Process Configuration

4.1 Introduction

In this chapter we explore the interactions between the market driver "customized products" and the supply chain processes (Figure 1.2 in Chapter 1). Specifically, we are interested in how the market drivers help determine the attributes of the product, and how the product attributes impact the supply chain processes. A product such as a car has a large number of attributes that include rate of acceleration, gas mileage, automatic transmission, and anti lock brake system (ABS), on which the customers express their preferences. In addition, car is defined by commercial attributes such as delivery time, quality, and price. Design engineers work with these attributes and determine the engineering specifications for mechanical and electrical components of the car. A product is therefore a bundle of attributes: customer specified, and engineering specified. The component designs along with the commercial attributes are then used to determine the supply chain processes that, for example, describe which supplier is going to build what, which production processes would be used, how much flexibility should be built in the chain, how to maximize component-commonality, how the component deliveries would be coordinated, how to obtain

	Premium service	Standard service
Market drivers	Customization, additional services, comfort, convenience	Price, acceptable service
Product attributes	Performance quality, flexibility, speed	Cost, conformance quality

Fig. 4.1 Airline service products.

scale economy, where and how new processes would be developed, and how to keep costs manageable. Thus, there are three distinct phases in product design: extracting the preferences of customers, mapping the preferences to engineering parts, and mapping the engineered parts to supply chain processes.

We revisit the airline example discussed in Chapter 1, and modify it for product configuration, as shown in Figure 4.1. Two market drivers, one focused at customization and the other on price, are being considered. The product-attribute bundle matching customization, comprise flexibility, speed, and performance quality. The airline calls this product the premium service. In a similar way the low-price market driver matches the bundle low cost, and conformance quality based on specifications. The airline calls this product the standard service.

Whether the airline would choose both products or pick just one of the two, would depend on the current capability of the airline, and whether it can retool itself quickly.

There are several issues that need to be addressed as a part of overall strategy. First, for a given set of customer preferences, the company needs to determine which attributes to include in the product. Second, as customers are not a monolithic entity, it may become necessary to create multiple market segments, one for each customer group. Third, customer preferences change in time, requiring redesign of products. Decreasing product life cycles or the demand for better performing products require rapid reconfiguration of products.

4.2   Customer Preferences

We explore the appropriate response to variations in customer preference in an incremental fashion. That is, we first discuss the mapping between customer preferences and product specifications, for a single product. We then study product designs for uncertain environments through different supply chain scenarios: risk hedging, responsive, and agile.

4.2.1 Mapping Customer Preferences

In general, the attributes valued by customers include performance, reliability, durability, appearance, and price, among others. Design engineers, on the other hand, refer to product features in terms of the components such as cylinders, valves, crank-shaft, gear box, transmission, and disc-brake. There obviously, exists a relationship between the engineering components of the product, and the customer desired attributes of the product. A map (or a linkage) in this context, is an "expression" of this relationship (abstract or simple). We first describe the notion of mapping customer desired attributes to product features, using quality function deployment. We then discuss some mathematical modeling approaches which have been found useful in capturing such mappings.

Quality function deployment is a set of relationships between customer needs, product components, and production needs. These relationships are expressed through one or more charts. Substantial cost savings in design, arising from the use of QFD, is reported by Hauser (1993). QFD can be deployed in several phases: product planning, parts configuration, process planning, and production planning. One or more charts may be used in each phase. Hauser and Clausing (1988), provide a special construct for such charts which they call the House of Quality (HOQ).

Design attributes are a set of parameters that can be used to describe engineering performance of a product. For example, performance parameters corresponding to the customer need "easy to operate", may include "time to perform the task", "degree of coordination required in task-performance", "on-line support available in performing the task", and "simplicity of task" (Hauser and Clausing, 1988). Note that a single design attribute may satisfy one or more customer needs.

Assume that $y_j = 1$, if the design feature j is included into the product bundle, and $y_j = 0$, if it is not. Thus, y_j is a binary variable. We define α_{ij} as the contribution of the design feature j in satisfying customer attribute i, w_i the relative importance of attribute i determined by customers, and c_j the marginal cost of design feature j. We capture the relationship between any two design features as r_{jk}. Let B denote the available budget. Therefore, corresponding to design decision y_j, customer satisfaction can be expressed as $\sum_{i\in I}\sum_{j\in J} w_i\alpha_{ij}y_j$. Thus, the mapping problem is expressed as

$$\text{Maximize } \sum_{j\in J}\sum_{i\in I} w_i\alpha_{ij}y_j = \sum_{j\in J} W_j y_j$$

$$\text{Subject to } \sum_{j\in J} c_j y_j \leq B$$

where $W_j = \sum_{i\in I} w_i\alpha_{ij}$.

To solve the optimization problem, simply rank order the product features by $W_j/c_j = (\sum_{i\in I} w_i\alpha_{ij})/c_j$, and include them in our choice set so long as the constraint $\sum_{j\in J} c_j y_j \leq B$ is not violated. It follows that design features that are strongly related to popular sets of customer preferences, and incur low unit cost to build, will be included in the optimal product bundle.

4.2.2   Supply Chain Scenarios

When customer preferences change rapidly, the attributes of the product need to do the same. There are two ways of managing this: introduce a stream of new products quickly as practiced by Zara, or create a platform for adding or replacing features from the bundle, with ease. The company may need to maintain a portfolio of products: functional products with customer preferences for stable attributes, and products with innovative attributes that may change rapidly. Innovative products are a special feature of a market driven supply chain. They share the characteristics of short life cycle, unpredictable and volatile demand, and high margins. Functional products, on the other hand, tend toward longer life cycle, standardized goods with predictable demand, and low margins. There are supply chain implications, as there needs to be a good fit between the supply processes and the products delivered to

	Functional Products	Innovative Products (demand uncertainty)
Stable supply process	*Efficient SC* Scale economy, Logistics efficiency, JIT	*Responsive SC* Modular products, supplier hub, postponement
Evolving supply process (supply uncertainty)	*Risk hedging SC* Inventory pooling, Information visibility	*Agile SC* Risk-hedging and responsive

Fig. 4.2 Types of supply chain. *Source*: Lee (2002).

customers. We can conceptualize different types of supply chains, using this notion of fit, as shown in the Figure 4.2.

Stable supply processes, where supply uncertainties are minimal, clearly support functional products. With functional products that emphasize low cost, supply chain can be made efficient through scale economies in production, inventory and logistics. A JIT system will also be easy to implement. The production system, as it would have relatively few changes, will have sufficient time for setups if production is postponed until demand becomes known.

A stable supply process can also support innovative products in several ways. Products can be designed in a modular form so that modules can be quickly replaced to create new products; product design can be postponed until the customer desired attributes are known; a supplier hub can be created, where designers may choose from a large variety of components for designing innovative products.

In environments with ever emerging new material, new processes, and new standards, or with suppliers that are vulnerable to disruptions, there will be considerable uncertainties in the supply process. Thus, while a functional product may face uncertainties only on the supply side, an innovative product would face uncertainties in both supply and demand. Information visibility is the best way of reducing uncertainties. When that is not possible, inventory pooling and capacity sharing would be some of the ways of creating hedges against supply uncertainty. Thus, companies would need to create hedges on both supply and demand sides — inventory and capacity pools, modular products, design postponement, and supplier hub.

4.2.3   Comparing Supply Chain Scenarios

We explore how optimal order quantities can be established. We assume demand is a random variable ξ with distribution $F(\xi)$. To capture supply uncertainty, we assume that for an order size of Q the supplier delivers a quantity γQ where γ is a random variable with support $[0,1]$. Retailer's unit sales price and unit cost are p and c, respectively.

For the responsive supply chain (with only demand volatility), we would have the standard news vendor solution,

$$F(Q) = (p - c)/p$$

For the risk hedging supply chain (only supply uncertainty), the retailer's profit for a known demand D would be

$$\pi = p\left\{ D \int_{\gamma Q \geq D} g(\gamma)d\gamma + \int_{\gamma Q \leq D} \gamma Q g(\gamma)d\gamma \right\} - cE(\gamma)Q$$

From the first order condition, we obtain optimal Q as

$$\int_{\gamma Q \geq D} \gamma g(\gamma)d\gamma = E(\gamma)(p - c)/p$$

For the agile supply chain (uncertain demand and supply), the retailer's profit would be

$$\pi = p\left\{ \int_{0 \leq \gamma \leq 1} \left(\int_{\xi \leq \gamma Q} \xi f(\xi)d\xi \right) g(\gamma)d\gamma \right.$$
$$\left. + \int_{0 \leq \gamma \leq 1} \gamma Q \left(\int_{\xi \geq \gamma Q} f(\xi)d\xi \right) g(\gamma)d\gamma \right\} - cE(\gamma)Q$$

As the profit function is concave in Q, we obtain optimal Q from the first order condition, as

$$\int_{0 \leq \gamma \leq 1} F(\gamma Q)\gamma g(\gamma)d\gamma = E(\gamma)(p - c)/p,$$

4.3 Product Re-design

Depending on the market drivers, a company may want to switch from producing functional products to producing innovative products. The company needs to address a number of issues, discussed above, for it to make such a transition.

4.3.1 Modular Design

A product can be designed as an aggregation of modules, each module comprising several components (Baldwin and Clark, 1997). Page and Rosenbaum (1987) discuss redesign of an existing food-processor product line at Sunbeam Appliance Company. They identified 12 modules of the product. Three of the 12 modules — motor, bowl, feed-tube — and two options for each module are shown in Figure 4.3.

For the example in Figure 4.3, it can be verified that there can be 32 variants of food processor (note that in Page and Rosenbaum's original example 69,984 variants were possible). Thus, if module i has k_i options and if there are L modules, the number of product variants would be $k_1, k_2, k_3, \ldots, k_L = \Pi_{i=1,L} k_i$. In the above example, $L = 5$ and $k_i = 2, \forall i$. Therefore, the maximum number of product-variants will be $32 (= 2^5)$. It can be verified that to create the 32 variants, each option will be chosen 16 times $(= 2^{L-1})$. In general, each option of module i will be chosen $\Pi_{\substack{u \neq i \\ u=1,L}} k_u = g_i$ times.

Assuming the cost of the jth option of the ith module to be c_{ij}, the cost of product variants is written as $\sum_i \sum_{j=1,k_i} g_i c_{ij}$. The decision to

Module		Option 1	Option 2
Motor		Regular	Heavy Duty
Bowl	Size	1.5 quarts	4 quarts
	Type	Regular	Side Discharge
	Shape	Cylindrical	Spherical
Feed Tube		Regular	Large

Fig. 4.3 Product variety.

produce M variants at a minimum cost is written as

$$\text{Minimize} \sum_{k_i} \sum_{i} \sum_{j=1,k_i} g_i c_{ij}$$

$$\text{Subject to} \prod_{i=1,L} k_i = M$$

4.3.2   Demand Uncertainty

We next study the impact of demand uncertainty. Assume a random demand ξ_v for variant v, with distribution $F(\cdot)$. Assume q_v units of variant v are produced. Note that the cost of production is written as $\sum_i \sum_{j=1,k_i} (\sum_{v=1,g} x_{ijv} q_v) c_{ij}$, where $x_{ijv} = 1$ if the jth option in module i is used in variant v and $g = k_i g_i$ for any i. Ignoring shortage cost, and assuming that leftover quantities of variant v can be salvaged at price s_v, the profit maximization problem would be

$$\text{Maximize}\,\pi = \sum_v [p_v E_{\xi_v}\{\min(\xi_v, q_v)\} + s_v E_{\xi_v \leq q_v}(q_v - \xi_v)^+]$$

$$- \sum_i \sum_{j=1,k_i} \left(\sum_{v=1,g} x_{ijv} q_v \right) c_{ij},$$

The module options must be built in advance of orders. Therefore, the number of options to build for each module must be decided in anticipation of orders. In real time, when a customer's preferences become known, appropriate module options are chosen from inventory to configure the product. It implies that ordering, order-processing, and the associated databases would require restructuring. The procurement process will need modification as well, because the set of vendors supplying components for a module-option must be managed jointly to avoid bottlenecks. As the rate of configuration of modules must track customer demand, only a subset of suppliers would be used at any time. In essence, the supply chain will be a virtual enterprise, with a demand driven supply network.

Lean design in a modular-design setting must exploit module and component commonalities that increase reusability. Obviously, module commonality minimizes the number of unique modules needed to create a given variety of products (by using common modules in multiple products). In the same vein, component commonality lets common components to be used in multiple modules. With common components, order size per component increases that enables better economy of scale for the vendors and hence lower component prices. In addition, the needed number of vendors decreases, enabling closer relationships.

HP printers used the notion of "core plus peripherals", which is a variation of modular design. Instead of creating L modules, they built a core printer with a large set of common modules that did not vary with customers. They also created peripheral components, each with specific functionality, and could be assembled to the core at customer sites. This allowed HP to postpone customization of printers for its European customers, who needed multiple language support. Lucent Technology used the same principle for successfully bidding on a Saudi contract. By building the core in advance of the order, they could satisfy the stringent delivery time constraint in the Saudi contract.

4.3.3 Design Response Time

In products such as food processor and car, modules are typically defined by engineers. In systems such as telecommunication, on the other hand, modules can be formed multiple ways by bundling components and supporting services together. Consider n components bundled into L modules ($L \leq n$), with n/L components per module. Assume t and $T(t \leq T)$ to be the time to assemble a component and a module, respectively. Clearly, the time to assemble an integral product (without modules) is nt, and the corresponding time for the modular product is $L(n/L)t + LT = nt + LT$. Thus, the total assembly time for the modular product ($= nt + LT$) is higher than that when it is not modularized ($= nt$). However, in a build-to-order mode, the assembly time *following* an order-receipt will only be LT for the modular product (as the modules would have been built in advance of the order), and nt for the integral product. It follows that modular design would enable a quicker response if $L \leq nt/T$. Note that for a given T, the BTO response time as well as product-variety increase in L. This reveals a tradeoff between product customization and quick response.

The next question is how to manage the transition of assembly processes from integral to modular (Lau et al., 2007). To make an integral product, components are assembled one at a time to the product, while the semi-finished product is moved from one work station to the next. To build a modular product, on the other hand, the assembly processes are divided into L cells, one for each module; and the components fed to the cells are organized by module type. Once the modules are completed they are configured as a product in the $L + 1$st cell. In a simple case with no sharing of work stations between modules, the current work stations may be *operationally* grouped into modular cells, without physically moving the work stations. Operational grouping would involve redefining the start of assembly at each work station, moving the semi-finished modules between work stations, and transferring the completed modules to the $L + 1$st cell. In a more complex setting where (say) cell i needs to use a work station "assigned to another cell", duplicate work stations may have to be created to overcome sharing conflict. It may be possible to use substitutable components (by redesigning them slightly) to avoid work station duplication.

4.4   Mass Customization

A major source of variability in demand is the need for customization (Zipkin, 2001). As discussed in Chapter 2, price incentives can influence demand variability. A different approach would be to keep demand as is, and have the customers manage their varying needs as in *adaptive customization*, or create service bundles for demand consolidation as in a *supplier consortium*. Collaborative customization can be implemented in both MTS (make to stock) and BTO (build to order) environments.

In MTS products are made in batches before demand is known. These products are designed using technology innovations so that customers may reconfigure them to different forms based on their needs. It is a common practice to engineer flexibilities in them so as to facilitate *adaptive customization*. Prime examples are adjustable seats in cars, food processors that can be quickly adapted for use as blenders and mixers (using a common motor for "drive"), sport-utility vehicles that can be transformed from a 2-wheel drive to a 4-wheel drive, thermostat control equipment on air-conditioning/heating systems that can be programmed for different temperature settings, and toys that can be transformed to different forms and shapes (e.g., from a monster to a race car). To endow products with such flexibilities, programmable technology comprising microprocessors and complex mechanical movements that respond to adaptation by customers, must be built-in in the product. It is clear that since customers do the adaptations themselves, only a limited number of platforms would suffice. Each product or platform will, however, have to be more complex, almost intelligent-like, to transform simple adaptation steps by the customer (to alter product characteristics) to a sequence of complex maneuvers.

Transparent customization, implemented in a BTO environment, tracks customer's purchases and uses a learning process such as artificial intelligence (AI) to reveal his/her preferences. Electronic commerce makes it possible to track customer purchases on-line. Amazon.com has taken it a step further where they use AI to discern patterns in customers' book reading habits. As new books emerge, the company matches attributes of individual books with customers' revealed reading preferences, and alerts customers to the new book, if a match exists. This is an effective way of neutralizing the negative effects of demand variability, as the company develops advanced intelligence on what the customer will be likely to order, and will have adequate time to get ready for it. This is another example of demand-shaping, discussed in Chapter 2.

Collaborative customization is realized when the manufacturer and the customer determine the product specifications jointly. When the customer of Matsushita Electric visits a showroom and tries out a flexible bicycle frame, specifications of the product are logged in automatically (Murakoshi, 1994). In a novel application of virtual reality, customers in a showroom put on headsets and gloves to "walk through" a kitchen and try out different layouts until satisfied, which is then recorded and used for order placement. The manufacturing system would experience extreme variability, as each order is unique. However, as the manufacturing equipment is reset automatically every time the system enters an order, programmed automation decreases variability that would be experienced otherwise.

Supplier consortium is used for customization, when a single supplier is unable to supply different needs (products) of the customer.

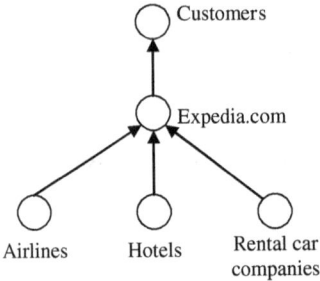

Fig. 4.4a Supplier consortium.   Fig. 4.4b Independent suppliers.

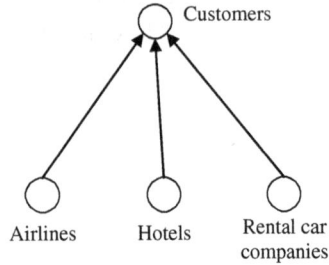

In the travel industry, customers routinely need airline tickets, hotel rooms, and rental cars. An intermediary such as Expedia.com creates bundles of these services to match the demand of different customers. In Figure 4.4a, Expedia operates as a consolidator for three types of services, and obtains volume discounts for customers for specific service bundles. Demand consolidation reduces variability that the vendor needs to handle. A variety of bundles is possible: one or more services, service type, regions of the country etc.

In Figure 4.4b the customer negotiates for services individually with each vendor.

4.5   Production Processes

A facility, and the processes within the facility, can be organized in many different ways: job shop, batch, product line, and manufacturing cell. The facility itself can be centralized, decentralized, or be a hybrid. Each of these process and facility architectures fits well with specific business frameworks (Chakravarty and Ghose, 1993). In a market driven environment, the facility architecture must respond to product categories, which we have studied previously as functional and innovative. Facilities can generally be categorized as centralized and decentralized.

4.5.1   Facility Configuration

While centralization conserves resources, decentralization of facilities, enabling locations closer to the markets, increases customization and decreases the cost of transportation of finished products. However, the cost of setting up facilities increases. The cost of transporting components and raw material to the facilities, if they cannot be acquired locally, may also increase. The decentralized facilities may be further categorized as product-specific, process-specific, and network, as shown in Figure 4.5. Note that in terms of the organization of processes, a product-specific setup is similar to a product line and a process-specific setup is similar to a job shop. In a product specific facility all processes of a single product, from fabrication to assembly, are performed in the same plant. A process specific facility, on the other hand, may specialize on a specific process such as the final assembly (of more than one product). In a decentralized network, as discussed in Chapter 3, the facilities may share resources.

Thus, with two product categories and four facility categories we can create eight variants of product-process interactions, as shown in Figure 4.5.

While a centralized facility for a functional product may make intuitive sense, decentralized facilities that are product specific and located close to markets can be justified if distribution cost and/or time are high. Similarly, process-specific facilities can be justified only

| | | Manufacturing plants | | |
| | | Decentralized | | |
Products	Centralized	Product specific	Process specific	Network
Functional	A single large facility producing for all markets. High cost of logistics.	Facilities producing specific products closer to markets, to reduce logistics cost	Final assembly in low-cost locations. Implies sub assembly shipments	Share inventory and other resources
Innovative	A large flexible facility producing a variety of products. High cost of logistics	Multiple facilities, each producing specific products at a location. Higher cost of facility	Flexible process that can switch to different products located in specific countries	Share process knowledge

Fig. 4.5 Product process interactions.

if process-expertise exists at those locations. The following facility-process architecture will be needed to support innovative products. For a centralized facility, it will need to possess flexibility to switch to different products. Decentralized product-specific facilities would require multiple facilities at any location, and process-specific facilities will have to possess flexible processes. Location of decentralized facilities will, obviously, depend on country-specific parameters such as cost, process knowledge, market size and supplier locations.

4.5.2   Restructuring Process Bundles

Restructuring of production processes becomes necessary whenever there is a transition from one cell in Figure 4.5 to another. Consider a centralized facility built for a functional product, sold in N marketing regions. A transition to product-specific facility would imply that N facilities would be built, one for each region. If identical products are sold in each of the N regions, the manufacturing processes will be the same in the facilities (100% overlap). Thus while the cost of building facilities would increase, the logistics cost and response time for fulfillment would decrease. The tradeoff between the increased facility-cost and the savings in logistics cost along with the gain in response time would determine how many product-specific facilities (not exceeding N) are built. A hybrid-architecture, with a lead facility ("centralized") located centrally and a number of satellite facilities (product-specific) closer to markets, may turn out to be the optimal. Implementation of such a hybrid can be hard, as the flows between facilities would require an exacting level of coordination for the system to be effective.

As an alternative, the manufacturing processes could be grouped into m non overlapping clusters, each cluster located in a region with the best process expertise and/or cost-advantage in that process-cluster. Such a facility is known as process-specific. Clearly, the output (semi-finished product) of one facility will be the input at another, until the finished product is delivered to the customer. One of the issues to be resolved is where to locate the final assembly. It can be located in a low-wage region for production-cost advantage, or close to a market for advantage in fulfillment cost and/or faster response.

If the supply is stable (low level of uncertainty), manufacturing emphasis will be on processes that run inexpensively. In an unstable supply environment, rapid-scalability of processes will be critical. The manufacturer can cover some of his risk by switching suppliers if need be. For this purpose, the manufacturer may create an online supplier-hub where supplier information is displayed. The manufacturer can now select new suppliers on the hub to compensate for the suppliers with unsatisfactory performance. In addition, the manufacturer may require the suppliers to assemble their components to the product at the manufacturer's facilities, and retain ownership of components until the assembled product is sold. The Volkswagen facility in Brazil is a good example (plants within a plant).

With innovative products, the need for a large variety, and an uncertain demand are the major issues. Modular design of products is the best response to the need for variety, as discussed earlier. However, transition to modular products will have major consequences for manufacturing and business processes. Assembly processes in the facility will need to be clustered and co-located. Procurement of components for specific module-options will require coordination to avoid bottlenecks. The suppliers of components of a module-option can work in a VMI mode through a lead supplier. If the supply base is uncertain, creation of a supplier hub and design of substitutable components should be considered, as discussed earlier.

4.6   Restructuring Processes in a Facility

Existing processes can be restructured by regrouping, re-sequencing, or by making them flexible. The company can split customer demand into stable and unstable components, discussed earlier. It can then build two facilities: a dedicated facility to produce to the stable part of demand by exploiting economy of scale, and an agile facility for the unstable component of demand. The National Bicycle Industrial Company (NBIC) has employed this strategy. Its customized bicycles (Panasonic), accounting for 2% of total production, contribute 27% of total revenues.

Process re-sequencing has been commonly used for delaying product differentiation (postponement). Benetton re-sequenced its operations by postponing dyeing of sweaters (to after knitting), and vastly improved response to demands for popular colors and reduce inventory of other colors. The dyeing process experience higher demand variability, as customers express their preference for sweaters in different colors. By delaying the dyeing task, Benetton did not have to commit quantities to different colors until the demand for colors was better understood. Benetton had to restructure its processes significantly though. The demand for knitting became smoother (demand consolidation). The process of knitting became simpler, as it did not have to be broken-up by colors. The dyeing process became more complex. Dyeing in bulk was not feasible, as dyeing of finished sweaters requires greater care.

Flexible machines can quickly switch cutting tools from an attached tool bank, and thus can respond the process changes. This implies that not only the tool banks are replenished with new tools dynamically, but also the movement of components to the machines (for processing) is controlled dynamically. Another approach would be to group the processes and traditional machines into manufacturing cells; the objective is to minimize the flow of work between cells. The manufacturing cells, unlike work stations, can handle a variety of products. These cells are also known as "plants within a plant".

4.7   Reconfigurable Manufacturing

One way of satisfying the varying needs of customers is to reconfigure the manufacturing system itself. This can be realized by reconfiguring the hardware and software components of the system that alter manufacturing capacity and/or capability (Kuzgunkaya and ElMaraghy, 2007). Hardware components that can be reconfigured are modular machine tools, material handling systems, and sensors. The software components are process plans, order processing, procurement, and system controls.

Reconfiguration of a production line to adjust its capabilities can be done in many ways: alter physical layout or modify the system controlling the movement of components. Work station layout can be altered if the assembly system is equipped with movable workbenches that can be adjusted according to customer needs. General Motor's Michigan Electronic Vehicle assembly facility is endowed with such flexibility, whereby, it can reconfigure to meet new demands; its throughput can vary from 2000 to 100000 cars per year. The other way would be to dynamically adjust the movement of components and assemblies so that they are not required to visit work stations in a linear fashion. Rockwell International created such capability in its flexible manufacturing facility in Milwaukee.

In redesigning a process, certain rules of common sense can help simplify the processes and reduce costs. These are elimination of redundant tasks, transforming sequential task to parallel tasks (if possible) so as to create a hybrid of sequential and parallel tasks, re-sequencing processes, reassigning tasks from specialists to versatile workers, centralizing knowledge management tasks, reducing input and output sources to minimize conflicts and confusion, and moving work to customers or suppliers (examples: customers entering own orders online, and ATM banking).

4.8   Adaptive Scheduling

In a dynamic manufacturing environment, uncertainties related to job arrivals, demand rates, due dates, and processing times do not permit strict adherence to production plans. A real time response is required to alleviate congestions and/or bottlenecks, which can be based on observed values of system states such as unfilled demand, resource availability, and strategic priorities. One way is to frequently update production schedules, which assigns jobs to machines. This is also known as dispatching. Chakravarty (1997) has framed real time scheduling as a pattern recognition problem, using data envelopment analysis (DEA).

In a manufacturing system, the typical input attributes for job scheduling would be number of parts in the system, machining time per part, and available machines. The typical scheduling rules would be the shortest processing time SPT, and expected due date EDD. Denote the jth observed value of input attribute of type i as a_{ij}. Similarly, denote the jth observed value of the output attribute (such as throughput), using the kth scheduling rule, as b_{kj}. The values of input and output attributes, in a "representative" observation, can be expressed as a convex combination (weighted sum) of previous observations (monotonicity property of DEA). We let λ_j to be the weight attached to the jth observation, where $\sum_j \lambda_j = 1$.

Consider the objective of choosing one of the two scheduling rules that would maximize throughput for new values of input attributes α_i, given the observed values. We let the binary variable $x_k (x_k = 0, 1)$ denote the choice of scheduling rule k. That is $x_1 = 1$ implies that SPT is chosen. Thus, the objective function to be maximized can be written as,

$$\text{Max } Z = \sum_{k \in K} \sum_{j \in J} b_{jk} \lambda_j x_k$$

The constraints, for "desirable" attributes (enhancing the objective function) for throughput, such as number of parts in the system and availability of machines, would be expressed as.

$$\sum_{j \in J} a_{ij} \lambda_j \leq \alpha_i$$

For the "undesirable" attributes for throughput (machining time per part), the constraint would be

$$\sum_{j \in J} a_{ij} \lambda_j \geq \alpha_i$$

The binary choice and convexity constraints would be

$$\sum_{k \in K} x_k = 1, \quad \text{and} \quad \sum_{j \in J} \lambda_j = 1$$

Note that the objective function is non linear. However, as the set of scheduling rules is not large, we may maximize $Z_r = \sum_{j \in J} b_{jk} \lambda_j x_r$ for $k = r$ by setting one of the $x_r = 1$, at a time, and $x_{k(k \neq r)} = 0$; and determine optimal k as $k = \arg\max_r (Z_r)$.

Chapter 5

Supplier Network

5.1 Introduction

The major decisions in managing a supply network are the design of the "physical" network, supplier selection, procurement, outsourcing, and logistics, as mentioned in Chapter 1. While these decisions are inter-related, there are unique features of each decision. While the network design creates a plan for which supplier can produce what and who can supply whom, other processes such as procurement and logistics create and implement contracts for purchasing and movement of goods. There would be many supplier options for each component, and the right set of suppliers must be chosen. As a component may be procured through multiple channels (e-markets, auctions, catalogs, retail stores etc.), the choice of channels would also influence the choice of suppliers. Together, these processes generate a large number of alternatives to choose from. In a market driven environment, with demand and supply uncertain-ties, the company needs to develop the capability to switch from one supply chain alternative to another. If demand is volatile it would pay to use a portfolio of suppliers who can individually satisfy different components of demand. If there is an increasing trend in demand, it would make sense to move away from general-purpose suppliers and select focused suppliers so as to exploit the scale economy.

The Hong Kong based Li and Fung has developed an efficient way to work with suppliers that possess overlapping capabilities. This enables the company to switch effortlessly between different garment-production projects. Thus, supplier relationship and outsourcing must be studied in a network context. Another example of being market driven is the Crocs Company. Crocs manufactures trendy shoes, and found a way to quickly mobilize its production capacity so as to produce multiple times during a selling season and adjust its production capacity to demand. Clearly, demand management discussed in Chapter 2, would need revisiting in the context of a network.

5.2   Supply Network Design

Consider a supply chain comprising three stages: component manufacture, assembly, and delivery to the customer. The supply chain needs to accommodate several types of components, each preferred by a segment of customers. The production and fulfillment options at each stage of the supply chain may be many. Assembly may be performed on a manual or automated line. Deliveries to customers may be made by air or overland. Thus, a supply chain with three components and two options each at assembly and fulfillment, will have $12 (3 \times 2 \times 2)$ alternatives to choose from.

Note that the quality of the delivered product is determined not only by the choice of components *but also by the assembly and delivery processes chosen*. Therefore, the product-as-delivered will correspond to a path in the supply chain. Assuming a customer prefers component A over B, air delivery over overland delivery, and automated assembly over manual assembly, he would be operating on a chain defined as "component A, automated line, air delivery". Because each design alternative of the supply chain corresponds to a delivered-product, a modified conjoint analysis can be employed to obtain customer preferences for all supply chain paths in the design, each path representing a product.

5.2.1 Supply Chain Design Alternatives

In general, a supply chain can be defined as comprising $n(i = 1, n)$ value-adding stages with $m_i(j = 1, m_i)$ options to choose from at stage i, so that the number of possible design alternatives for the supply chain would be $\prod_{i=1,n} m_i$, some of which may not be infeasible. Assume that $K(k \in K)$ is the set of feasible design alternatives (paths in the network). It is clear that some of the m_i options available in stage i would not be used in the supply chain set K, if it is not cost effective to do so. We denote option j at stage i by the pair ij, and the set of chosen options in stage i as $S_i(j \in S_i)$ and $S = \sum_{i=1,n} S_i$. Clearly, $S \subseteq T$, where T is the set of all ij pairs. We assume that customers are capable of assessing the utility of option ij in the context of the delivered product, as discussed earlier. For example, a customer may prefer cars with anti-lock brakes to those with regular brakes, and air delivery to overland delivery, if price remains the same.

Assume that the customer preference for option ij chosen in design alternative k is expressed as u_{ijk} (say, on a 0 to 10 scale), and W is the market size (the number of potential customers). Using the Logit model, we may express the demand for option j in stage i as $D_{ijk} = \frac{W v_{ijk}}{v_C + \sum_{i=1,n} \sum_{j \in S_i} \sum_{k \in K} v_{ijk}}$, where $v_{ijk} = Ln(u_{ijk})$ and v_C represents the corresponding preference score of the competitor's supply chain (Chakravarty and Blakrishnan, 2001). Assuming the price of delivered-product on supply chain k to be p_k, and the total cost of component options used on product k to be c_k, the total profit for the supply chain design is expressed as

$$\pi_S = \frac{W \sum_{k \in K} (p_k - c_k) \left(\sum_{i=1,n} \sum_{j \in S_i} v_{ijk} \right)}{v_C + \sum_{k \in K} \sum_{i=1,n} \sum_{j \in S_i} v_{ijk}},$$

We rewrite the profit function as $\pi_S = \frac{X_S}{v_C + Y_S}$, where $X_S = \sum_{i=1,n} \sum_{j \in S_i} x_{ij}$, $Y_S = \sum_{i=1,n} \sum_{j \in S_i} y_{ij}$, and $x_{ij} = W \sum_{k \in K} (p_k - c_k) v_{ijk}$, $y_{ij} = \sum_{k \in K} v_{ijk}$. Clearly, $\pi_S = \text{Argmax}_{L \subseteq T}(\pi_L)$. Chakravarty and Blakrishnan (2001) establish that as π_S is linear in X_S and convex in Y_S, it is the case that $ij \in S$ and $\ell\lambda \notin S \Rightarrow x_{ij}/y_{ij} \geq x_{\ell\lambda}/y_{\ell\lambda}$. Therefore to obtain optimal S, keep adding options (ij pairs) to S in descending order of x_{ij}/y_{ij} as long as π_S increases; stop, if π_S starts to decrease. For a given S, the optimal set $R(R \subseteq K)$ of supply chain alternatives (paths) can now be established.

In the optimal design discussed above, it is possible that some ij links, such as air-delivery, would have higher demand than other links on the same path in the supply chain. If the current capacity of the ij link is not adequate, the capacity constraints must be incorporated in supply chain design.

5.2.2 Demand Volatility

We can capture demand volatility by adding a random component θ to the demand expression, so that $D'_{ijk} = D_{ijk} + \theta$. For example, there may be uncertainty in the preference ranking of components (by customers), causing the demand of a delivered-product to be uncertain. Using a news vendor type analysis, we can easily establish that the cardinality of the set R will be larger. That is, demand volatility would cause "excess" supply chain alternatives to be in the design set.

In addition to revenue and cost associated with each supply chain alternative, there may be other customer preferred attributes, such as fulfillment time of the delivered product and its quality. For example, while buying components from suppliers would be less expensive than inhouse production, it may take longer. In some cases, we may want to choose the minimum cost supply chain alternative for a given response time. The response time constraint would decrease the size of the feasible set of alternatives we need to consider.

5.3 Cost of Building Relationship

There will also be a fixed cost for building relationship with a chosen supplier, in addition to purchase cost. This will be a critical cost to consider when there are multiple suppliers per component. The capabilities of individual vendors (in terms of component options they can supply) implicitly determine the minimum number of vendors that must be used. While it may be beneficial from a fixed-cost (overheads) perspective to reduce the number of vendors, the chosen set of vendors, together, must be able to deliver all desired components. While a multi-purpose (versatile) vendor is capable of supplying a larger set of component options, it would be at higher unit costs than a special purpose (scale economy) vendor who supplies fewer options. Therefore, it would be necessary to identify the group of vendors that, together, are synergistic with the market drivers (Chakravarty and Blakrishnan, 2008).

As an example, assume supplier s_1 can supply components 1 and 2, supplier s_2 can supply components 1 and 3, and supplier s_3 can supply component 3. We need to consider seven supplier choice scenarios: (s_1), (s_2), (s_3), (s_1, s_2), (s_2, s_3), (s_3, s_1), and (s_1, s_2, s_3). Note that for the scenario (s_1, s_2), component 1 may be bought from s_1 or s_2. Similarly, option (s_2, s_3) implies that while component 2 will not be available, there would be two suppliers for component 3. Thus, for a given supplier choice scenario, we may determine the minimum cost of an option pair ij, if it is available. Using this cost we may determine optimal set of options, S, (as before). We can then proceed to find the supplier scenario that would generate the "global" optimum. Further, one can then assess whether it makes business sense to help this group of suppliers *develop new capabilities so that the company may exploit new opportunities* in the market place.

5.3.1 Consortium

The supply chain will be more resilient to disruptions, if there are multiple suppliers per component. To exploit this resiliency, we would generate optimal option set S for all supplier scenarios. If one or more suppliers are disabled, the contingency plan would trigger the next best supplier scenario and the corresponding option set S. However, switching from one supplier scenario to another would require that new relationships be developed. It would be prudent to develop such relationship, before contingencies arise. For example, each supplier scenario may be conceived of as a supplier consortium. One way of managing such a consortium would be to develop a lead supplier who would be responsible for coordinating other suppliers in the consortium. It would also make sense to alter product design to a modular form, with one module assigned to a consortium. The lead supplier can now be made responsible for acquiring components from other suppliers in the consortium and assembling them into modules. The success of the consortium, to a large degree, would depend on the ability of the lead supplier to monitor the production capacity and process capability of consortium members. Clearly, information transparency within the consortium coupled with appropriate incentives would be critical.

5.3.2   Monitoring Suppliers

Establishing a continuing bilateral relationship requires stringent screening of other parties in a relationship. Therefore, supplier-monitoring is an important step in ensuring supplier's ability to perform as expected. It may range from financial checks to examining a supplier's manufacturing operations, production capacity, personnel, and technological capabilities. As financial data may be confidential, suppliers may be asked to submit financial statements, balance sheets, and cash flow, to a third-party (such as an accounting firm) with a confidentiality agreement. The parties agree that the accounting firm would provide the manufacturer with a scorecard for each supplier, based on financial ratios. To ensure supplier participation, award of new business can be made contingent on their submitting required data. Such information helps the manufacturer identify suppliers that might be in trouble or headed for trouble.

Another useful strategy is to study suppliers' performance metrics, based on data on delivery, quality, and agility. With this data it would be easy to create monthly supplier scorecards to assess supplier's future delivery and quality trends.

5.4   Procurement

Kraljic (1983) suggests that purchased items be categorized into four groups based on profit potential and supply risk: leveraged, non critical, bottleneck, and strategic. Leveraged items, with high profits and low supply risk, permit the company to implement targeted pricing, product substitution, and negotiation. Non-critical items, with low profits and low supply risk, are amenable to product standardization, volume discounts, and efficient transaction-processing. Bottleneck items, with low profits and high supply risk, require supplier management through volume insurance, vendor monitoring, and backup plans. Strategic items, with high profits and high supply risk, require careful analysis for forecast accuracy, supplier relationship, make/buy tradeoff, purchase staggering, risk analysis, and contingency planning.

5.4.1 Procurement Strategy

In a market driven environment, we need to consider purchasing strategies based on three dimensions: demand volatility, supply uncertainty, and purchase cost. While for low demand volatility scenario Kraljic's framework would be appropriate, we would need to modify the framework for a *high demand volatility*. For each of the four scenarios shown in Figure 5.1, we discuss the implications of high demand volatility.

Scenario A: High unit cost and stable supply

With a stable supply, the supplier would be receptive to changes in the procurement system. With a high cost of purchase and high demand volatility, the manufacturer would gain by postponing purchase of components to a time when demand becomes known. Clearly, purchase postponement implies shifting inventories from the manufacturer's assembly plants to the suppliers. Suppliers may need incentives such as increased price for holding inventory. This can be studied as a special case of the consignment inventory concept. Purchase postponement would also require modifications to the logistics system, which would need the capability to transport large quantities as and when needed.

	Low supply uncertainty	High supply uncertainty
High purchase cost	High unit cost Stable supply	High unit cost Unstable supply
Low purchase cost	Low unit cost Stable supply	Low unit cost Unstable supply

Fig. 5.1 Purchasing strategy.

The manufacturer may prefer a buy-back contract to share the risk of unused components with the supplier. Consignment inventory, where the supplier owns the inventory of components until the time they are used by the manufacturer in assemblies, would provide protection against demand volatility. The manufacturer may also use hedges such as the option-to-purchase at an agreed upon price that would be exercised at some future date when demand is high.

Scenario B: High unit cost and unstable supply

With an unstable supply, the manufacturer would insist on a delivery contract with the supplier. The manufacturer would pursue risk diversification by earmarking multiple suppliers for the same component, using spot markets, and purchasing online from electronic marketplaces. Vendor managed inventory, where the supplier is responsible for replenishments to ensure an agreed upon service level, would be in the manufacturer's interest. The manufacturer would also need to pursue spot market purchases more aggressively. With high supply uncertainty, the supplier is unlikely to agree to either consignment inventory or buy-back contract.

In both A and B scenarios, the manufacturer may pursue JIT purchasing. This can be achieved by developing close relationship with the supplier that enables information transparency between the parties. The supplier, for example, could be involved in the manufacturer's product design and market research teams. The supplier will have advance information not only on the type of components the manufacturer is likely to order but also on possible order-size. The manufacturer has visibility of the supplier's inventory and his component production plans, and it helps him decide which orders, and how much of it, to direct to particular suppliers.

Scenario C: Low unit cost and stable supply

With a stable supply and demand volatility, it would be optimal to order in large lots and build inventory. Although it would be feasible to implement VMI and consignment inventory, the value of it for low cost items would only be marginal. A possible approach would be to commit to a dollar value of purchase. Order details such as product type, product quantities, and delivery time are then communicated when the demand becomes known. Although the supplier faces uncertainties in component type and quantity, he safeguards a business volume for the time of contract.

Scenario D: Low unit cost and unstable supply

With low purchase prices and high demand volatility, manufacturer's optimal strategy would be to build a large inventory. However, with a large supply uncertainty the manufacturer would need to diversify risk with purchases from multiple suppliers.

5.5 Restructuring Outsourcing

Outsourcing can create competitive advantage for the company for several reasons. First, the company can take advantage of the partner's scale economy. This is so because the outsourcing partner would have other customers and, hence, it can bring down its unit cost by aggregating orders into large volume for production. Thus, the partner would specialize in a limited set of tasks and sharpen its expertise through innovation and learning. Second, the company can retain agility, as its investment in resources will be relatively low, having outsourced the tasks. Third, it can get the outsourcing partners to quickly fulfill customer orders through drop-shipping. Dell, for example, has used UPS to pick up monitors and CPU from Dell's contract manufacturer, match them according to customer orders, and deliver them. The company may also arrange its outsourcing partner to set up a plant (such as final assembly) close to the company's customers, thus leveraging the partner to implement its postponement project. Cisco and FedEx have relied on a similar arrangement.

In any outsourcing practice there may be loss of knowledge, as the principal may not be involved in performing those tasks. This causes a degradation of critical capabilities, which the company may find very hard to rebuild. A related issue is the over dependence on the supplier causing a lock-in. It will be hard to switch out once lock-in happens, leading to economic inefficiency. Second, the task related information shared with the supplier may be leaked to competitors, leading to competitive disadvantage. Third, as the company may not be fully able to monitor the performance of the outsourced tasks, the company may lose control of how the task is completed. Fourth, there may be coordination and expediting problems if the terms of agreement between the partners are ambiguous. Fifth, poorly designed contractual agreement can lead to unnecessary litigation. Sixth, outsourcing, especially to offshore locations, can make the supply chain more vulnerable to disruptions due to distances involved and reduced transparency between the partners (information asymmetry).

5.5.1   Scope and Importance

The strategic value of outsourcing can be understood in terms of the scope and importance of outsourcing. Scope can be defined in terms of the nature of relationship between the two parties: transactional, shared responsibility, and supplier leadership. If the outsourced tasks are very specific and standardized, the instructions for suppliers can be easily codified and transmitted. An arms-length transactional relationship with the suppliers would now be practical as well as sufficient, though there would be a large number of qualified suppliers to manage. In contrast, if the tasks are somewhat complex, information exchange would need to be a two-way process. This would call for shared responsibility for task transference, monitoring, expediting, and fulfillment. Outsourcing a very complex task, such as design of a new product, may require the supplier to assume full responsibility for design, prototyping, testing, and market-positioning. It will also require the supplier to coordinate the procurement of components and material from other suppliers. This type of outsourcing relationship will have to be based on a great deal of trust between the parties.

Importance of outsourced tasks in terms of the competitive advantage of a company can be categorized as low, medium, and high. A car manufacturer may consider the engine to be of high importance, brake system of medium importance, and electrical systems of low importance. The underlying issues in this context are knowledge and production capacity, as discussed in Chapter 4. The company may possess both design-knowledge and production capacity for engines, it may possess design-knowledge but no capacity for the brake system, and possess neither design-knowledge nor capacity for electronics. A

Importance

	Low (Non core)	Medium	High (Core)
Full outsourcing	**Contracts**		**Alliance** (Walmart and P&G)
Shared responsibility		**Partners** (JIT)	
Specific tasks	**Transactions** (Spot market)		

Scope (label at left of the table, aligned with "Shared responsibility")

Fig. 5.2 Outsourcing strategy.

possible outsourcing portfolio would be for the company to design and make engines inhouse, design brake system inhouse and outsource its manufacturing, and outsource both design and manufacturing of electronics.

The interaction between scope and importance of outsourcing is shown in Figure 5.2.

5.6 Offshore Outsourcing

Offshore outsourcing spending of US based companies has increased from $5.5 billion in 2000 to more than $17.6 billion in 2009. It creates competitive advantage by leveraging operations efficiency at remote locations through advanced technology so that a multi-shore delivery system can work as a single virtual unit, regardless of where the operations are actually located. This results in very competitive economies of scale for the offshore facility in satisfying customer orders in the client's home countries, which can be passed on to customers. On the downside, it may expose the company to higher levels of uncertainty. The company must decide whether to keep manufacturing in-house or contract with global suppliers and, if so, to what extent.

The Griffin Manufacturing, an apparel supplier in US, faced the prospect of losing manufacturing to low cost producers in Honduras. However, because of the extended logistics line to Honduras, orders from the customer (a design house in US) had to be placed 8 months prior to the start of the selling season, and had to be based on long-term forecasts (Stratton and Warburton, 2003). In addition, the Honduran suppliers lacked the capability to respond quickly to changes in demand in real time. During the selling season, actual orders came with very short delivery windows and, therefore, orders that deviated from forecasts resulted in losses from markdowns and/or lost sales. Griffin sensed an opportunity and positioned itself in a manufacturing niche, with quick response capability. The customer saw the benefit in splitting her orders into two: an early order for the forecast-amount of "basic" products (at low price) with the Honduran suppliers, and a second order after the start of the selling session with Griffin to make up the variations from forecast (at a premium price).

The major dimensions in such decisions are integration (insource or outsource) and location (onshore or offshore) of servicing facilities. We denote a domestic facility as onshore and a foreign facility as offshore, as shown in Figure 5.3. Metters (2008) uses a similar framework to discuss the typology of offshoring.

There are obvious tradeoffs between cost, quick response, and risk in moving from onshore to offshore manufacturing or services (Figure 5.3). It suggests choices for firms in terms of various combinations of onshore and offshore facilities, as in the case of Griffin. Another example is

	Onshore	Offshore
Insource	**Domestic insourcing** Full control Complex products Quick response High cost	**Offshore insourcing** Usually low input cost Unknown territory, Delayed response High operations risk
Outsource	**Domestic outsourcing** Low investment risk Low cost from scale Low operations risk Simple products Quick response	**Offshore outsourcing** Low investment risk Low input cost Low operations risk Delayed response Uncertain yield, Quality concerns

Fig. 5.3 Offshore outsourcing choices.

EDS's portfolio of onshore and offshore facilities — based on uncertainty, cost, and language requirements (Kalakota and Robinson, 2004). Such a portfolio is essentially a choice of cells (one or more) in Figure 5.3 where the firm may wish to position itself. We next describe the major operational characteristics in the two offshore cells.

5.6.1 Operational Characteristics

A dedicated offshore center is owned and managed by the supplier and provides components to the onshore manufacturer. The manufacturer seeks to minimize her cost by leveraging the knowledge of the established supplier who possesses infrastructure and expertise to build and manage a facility. The onshore manufacturer gets the benefit of outsourcing by limiting the cost of investment in an uncertain environment. In contrast, a captive center is owned by the onshore manufacturer. An owner-operated facility becomes desirable when offshore operations are critical and/or proprietary information must be safeguarded. However, to operate such a facility, the onshore manufacturer would have to train and retain employees, and bear the risk of a high start-up cost.

To minimize her exposure to offshore operations risk, the onshore manufacturer may opt to acquire the facility only if the facility is sufficiently productive. This is known as the build-operate-transfer (BOT) model (Chakravarty, 2009a). The facility is transformed from a dedicated offshore center to a captive center. A firm that stands to benefit from a captive center but lacks business experience in a specific region of the world would find BOT appealing. An example is J.D. Edwards (Kalakota and Robinson, 2004). JDE reduced its cost of offshore operation by leveraging the infrastructure of Covansys, a technology services company in India. JDE (the principal) signed a six-year contract with Covansys with the stipulation that JDE could assume ownership of the facility at the end of the contractual term, or walk away. A related issue, therefore, is whether the onshore manufacturer should make the acquisition of facility contingent on the state of productivity. In addition, the manufacturer may consider providing incentives in order to motivate the offshore supplier to upgrade plant productivity.

5.6.2   Capacity Strategy

Both demand and supply uncertainties would play roles in determining the capacity-strategies of the onshore manufacturer and her offshore supplier. The onshore manufacturer would be interested in acquiring capacity for manufacturing and would consider using a portfolio of onshore and offshore facilities. For example, the manufacturer may maintain an onshore presence for rapidly responding to demand-changes, while exploiting the offshore cost advantage, through a BOT arrangement.

As shown in Figure 5.4, in a market driven environment, the source of volatility may reside in a domestic or a foreign market. For a volatile domestic market, the supply chain response should be as in the case of Griffin Manufacturing discussed earlier. However, if the source of market volatility is in an important foreign market, an appropriate supply chain strategy would be to redesign the product in a modular fashion with a large core and add-on peripherals for customization. If quick-response is the market driver, the generic core can be produced in an onshore facility (in advance of demand realization) and shipped to the offshore facility where peripherals can be added quickly, based on actual demand. Lucent Technologies adopted this strategy to win a Saudi order, which had a 3-week delivery lead time from the time the contract was awarded. HP used such a postponement strategy to add peripherals to its core printer at a distribution center in Europe; the objective was to customize the printer to the different needs of European customers (language, power supply etc.)

	Domestic market response	Foreign market response
Onshore facility	Reactive capacity (Griffin)	Generic core (inhouse)
Offshore facility	Speculative capacity (Honduras)	Customized add-ons (outsource?)

Fig. 5.4 Market driven onshore/offshore capacity portfolio.

5.7 Supplier Relationship Management

Relationships can be considered to be of four different types according to Himmelman (1993): networking, coordination, cooperation, and collaboration. Networking is the exchange of information for mutual benefit. Coordination is driven by a common purpose that would create mutual benefit through exchange of information and altering of activities. Cooperation is driven by a common purpose that would create mutual benefit through exchange of information, altering of activities, and resource sharing. Collaboration is driven by a common purpose that would create mutual benefit and enhance partner's capability through exchange of information, altering of activities, and resources sharing. Supplier facing activities such as sourcing, procurement, logistics, technology transfer, and sharing of best practices, require good relationship.

Hartman et al. (2001) explain how the major characteristics of a relationship can be defined in terms of trust, commitment, involvement, and satisfaction. Trust is the basis of any relationship and can be describes as the degree of confidence in how the partner would behave. While trust is not a major factor in transactional relationships, it becomes critical in a strategic relationship, such as developing a *new* product with a partner, where formal contract documents cannot cover all eventualities. Commitment defines attitude towards a long term relationship. Toyota's business practice is based on making long term commitments to its suppliers. This commitment stipulates continuing business relationship, while learning continuously to improve performance. This policy has been very effective in developing small suppliers in Japan. Involvement can be defined as getting to know in each other's business practices. This provides an opportunity to "dirty your hands" and learn from each other's experiences. Formally stated, involvement is a mechanism for knowledge transfer in an ongoing basis. The last component of relationship is satisfaction with partner's performance. It is a proxy for the effort the partner makes relative to what is expected. The challenge would be how to integrate satisfaction expectation with commitment. Clearly, expressing a lack of satisfaction should not be construed as a lack of commitment, and commitment should not be allowed to promote opportunistic behavior by the partner.

As partners may possess varied degrees of sophistication and as relationship with all of them may not be equally valuable to the company, the optimal strategy would be to think in terms of a portfolio of relationships. For example, to a manufacturer, collaboration with suppliers and retailers would be of more value than that with banks and freight-agents (Clark and McKenney, 1994). Such relationships can be conceptualized as exploratory, occasional, significant, and intensive. The explorative format is a pre-relationship exchange of ideas. An occasional relationship signifies a low frequency of interaction, and it would be adequate if the partners were only interested in access and information visibility. A relationship is called significant when it requires interfacing of demand with the supply chain processes. Finally an intensive relationship indicates joint decisions in a project management environment. This enables the partners in coordinating their activities and updating real time resource decisions.

The company would like to decide in which relationships to invest so as to build the portfolio of relationships. The two major determinants in such decisions are the value and cost of collaboration. The factors that impact value creation and costs in any relationship are the frequency of interactions with a partner, revenue and cost per interaction, and the cost of setting up a relationship. If a partner is technologically unsophisticated but can create a significant value in the supply chain, the manufacturer may wish to subsidize the partner for it to become technology savvy. The value (and cost) of relationship would obviously depend on the frequency of interactions between the parties — higher the frequency, higher the value and the cost.

For coordination of supply chain processes in a relationship, partners need to agree on the purpose of coordination and means of achieving it that may necessitate certain changes in their operations. Cisco Systems coordinates supply and demand across its supply chain with an intelligent-software for early detection of conflicts and their resolutions (Lee and Whang, 2001). Coordination should be tailored to specific business processes such as, order placement, order fulfillment, logistics, and payment. Exchange of information helps coordination in multiple ways: selecting suppliers based on their capability and availability, negotiating commitment and remunerations, coordinating service delivery, and resolving execution problems. Information visibility reduces the negative effect of upstream amplification of demand variance (bull-whip) — from the customer to the supplier through the retailer, wholesaler, and manufacturer (Lee et al., 1997).

Collaborative Planning Forecasting and Replenishment (CPFR) is based on the idea of setting up rules for resolving conflicts in forecasts (Aviv, 2001), and strengthening partner's operations. It has built in metrics for performance evaluation, accountability, and corrective actions (VICS, 2001). Workflow (Basu and Kumar, 2002) may be used for crafting collaboration. It can track a supplier's production related events to keep the customer informed. If the customer is concerned of supplier's ability to deliver an order on time, he/she may run the electronic workflow that links the ordering event to a set of order fulfillment processes of the supplier. Simultaneously, the workflow updates all documents to ensure that the customer has the most recent information at all times. Collaboration must allow for aligning of the incentives (of partners) with overall objectives. This requires an equitable sharing of both gains and risks among the members.

5.8 Contracts

A contract is a formalization of the terms of agreement between two partners in a supply chain. It usually specifies how the partner is expected to perform. In some contracts, it also specifies the corrective actions if the partner's performance is below par. Contracts that are based on measurable values such as quantity, time, and price, are easy to implement. In contrast, contracts based on partner-specific values such as cost and effort, suffer from information asymmetry and are hard to implement.

Supply chain contracts can be of three types: cost-plus, fixed price, and incentives (Bolton and Dewatripoint, 2004). Many procurement contracts are in fact a combination of the three, specifying incentives on some aspects and fixed-price on others. The contracting environment determines how contracts may be combined. While a contract is a convenient way of moderating partner behavior, it is of little value without a mechanism for ensuring conformance.

In a cost-plus contract the buyer agrees to pay a certain percentage above the supplier's actual unit cost. This insulates the supplier against all cost escalations, and provides little motivation for him to improve productivity. A special form of cost-plus is the capped price contract, which specifies an upper limit for the unit cost. Such contracts are usually used for professional and consultancy services. If the actual cost increased beyond the cap, the supplier could make a case to increase the cap. The rationale behind such a contract is that for projects such as R&D the buyer is unable to estimate how much work is needed. The cost plus contract would benefit the buyer in scenarios where production cost decreases due to advent of new technology or new processes. In situations where quality is non-verifiable, the cost plus contract has the advantage that it does not motivate quality cut-back by the contractor to save production cost.

In a fixed-price contract, price can be determined through a bidding process (reverse auction). The supplier bids a fixed price but must satisfy buyer stipulated specifications. While the supplier is not rewarded for improving quality, he may be penalized if quality is below the agreed standard. In this case, the supplier bears all the risk associated with cost escalation; an increase in cost decreases the supplier's profit. Therefore, while the supplier is motivated to reduce cost, fixed prices in the contract diminish the supplier's incentive to do so. In addition, the supplier would not be motivated in developing and maintaining good relationship with the buyer. Such contracts can be effective if (i) products are not complex, and (ii) markets do not suffer from unexpected shocks. In certain variants, contracts may include adjustments for fluctuations in inputs prices. Such adjustments are typically linked to a national price index, but the supplier still bears the risk of fluctuations in production cost.

Between the two extremes of cost plus and fixed price lie incentive contracts that typically include targets for cost and profit. Such a contract specifies a profit adjustment mechanism that ensures that losses from cost overruns are shared by the two parties. While the amount of profit or fee payable under the contract is related to the contractor's

Performance the contractor is never left entirely to bear the risk of fluctuations in production costs.

Incentive contracts have been extensively used in complex procurements in industries such as the construction and defense industries. A cost incentive contract can be expressed as $p = c_t + \alpha(c_a - c_t) + m = \alpha c_a + (1 - \alpha)c_t + m$, where p is the price charged by the supplier, c_a and c_t are the actual and target costs respectively, m is the supplier's margin, and α is the cost sharing parameter. Note that the incentive scheme is implemented through the sharing parameter α. If the actual cost is below the targeted cost, because of productivity improvements, the buyer benefits by a fraction α of the cost saving. Similarly, if the actual cost exceeds the target buyer shares the same fraction of cost overrun.

The cost sharing parameter α plays a crucial role on the supplier's incentives to decrease cost. A high value of α acts as a negative incentive for the supplier, as it causes a large fraction of a cost overrun $\alpha(c_a - c_t)$ to be passed on to the buyer, *reducing* supplier's motivation to decrease cost overrun, or to increase cost saving. Thus, the power of the incentive scheme decreases in the value of the cost sharing parameter α.

Choice of α is influenced by the supplier's predisposition to fluctuations in income $(p - c_a)$, that is, the degree of risk aversion. The choice is also affected by extent of uncertainty in controlling cost, and the investment and effort required in reducing cost. The supplier would prefer a low value of α if his risk aversion to income fluctuations is high, or the uncertainty in controlling cost is low, or the investment needed for cost reduction is high.

Incentive contracts can be designed for additional performance attributes, such as percentage of shipments delivered on time, as $p = c_t + \alpha(c_a - c_t) + \beta(r_a - r_t) + m$, where r_a and r_t are the actual and targeted performances respectively, $\beta(r_a - r_t)$ represents the bonus payment for good performance.

In most contractual environments the supplier bears the brunt of the endeavor in providing precise accounting measures of realized production costs, while the buyer has to measure quality levels. If the cost of implementing a contract is high, the buyer may adopt a contract such as the fixed-price contract that is easier and less costly to manage. While a fixed price contract shifts the procurement risk to the supplier, it allows the buyer to ignore possible discrepancies between estimated and realized productions costs, and requires little information for implementation.

Lack of reliable accounting measures may further undermine the effectiveness of incentive contracts, and result in choosing fixed-price contracts. Information asymmetry plays a big part in whether a contract can be implemented successfully. Note, for example, automotive vendors have been very reluctant to participate in improvement initiatives that would require them to share cost data with the manufacturers for fear that their margins would be squeezed.

In a profit sharing contract, the parties agree to share the total profit, measured as the buyer's revenue less the supplier's cost. It eliminates the negative effect of double marginalization and it coordinates the supply chain. However, the profit sharing proportion is subject to negotiation and provides advantage to the party with more power. Buy back contracts, where unsold items can be returned by the retailer at a discounted price, leads to sharing of demand risk by both retailer and manufacturer. As in profit sharing, negotiation is required to set the amount of discount on the items returned.

Contracts can be designed to make the retailer (agent) responsive to high demand volatility in the manufacturer's (principal) product. This can be done by basing the payment on actual sales of the retailer, instead of the quantity purchased from the manufacture. Consider, for example, the case of Whirlpool who uses Sears to sell its washing machines. If Whirlpool doesn't offer lucrative margins, Sears may motivate shoppers to buy its private-label brand, Kenmore. To overcome this problem, Whirlpool created a reward incentive based on the actual sales of their products at Sears' locations. Another example is the baker who concerned about demand volatility rewarded his managers a fixed amount for keeping its shelves filled so as to minimize shortages. This provided no incentives for the managers to increase sales, as the best way to keep the shelves full was to cancel all orders! In addition, there was no satisfactory way of observing the effort put in by the managers. The baker redesigned the incentive by rewarding the managers based on the value of actual sales.

In a market driven environment, the contracts must reflect the uncertainties in the marketplace. This can be achieved in multiple ways. The buyer can incorporate a delivery window instead of a fixed delivery time, permit a small variation in the quantity delivered, permit the wholesale price to vary within some bounds, and buy insurance contracts or financial hedges to insulate oneself against market volatility. Examples of creative risk sharing contracts would be, capacity reservation, build-operate-transfer, forming supplier consortium, and forecast revision.

Chapter 6

Managing Disruption Response

6.1 Introduction

In this chapter we explore the impact of the market driver called disruptions, as described in Chapter 1 (Figure 1.2). Disruptions in a supply chain can be caused by several factors including terrorism, infectious diseases, natural disasters, and human errors — all potentially catastrophic to the national economy and human life. The likelihood and impact of such disruptions are hard to predict and measure. The impact of natural disasters can be huge — the economic loss from hurricane Katrina is estimated to exceed $125 billion (Finkle, 2005). Similarly, the loss from the terrorist attack on world trade center in New York (September 11, 2001) is estimated to exceed $100 billion. There have also been huge losses from human errors — BP oil spill (2010), Chernobyl nuclear disaster (1986), and Bhopal chemical leak (1984). In some cases losses to individual companies have been high — Ericsson lost over $400 million when fire-damage forced closure of one of its suppliers (Rice and Caniato, 2003).

The uncertainties associated with disasters create complexities in response decisions, as disasters are typically characterized by a surge in demand. Although the social and economic consequences of inadequate or delayed relief can be catastrophic, the appropriate level of relief effort must depend on a multilateral tradeoff between disaster severity, infrastructure disruptions, demand surge, cost of redundancies, and social cost. The relief providers must also work in collaboration with the retailers of supplies, and infrastructure providers. We explore how the supply chain processes such as procurement and fulfillment are impacted, and how they may be reorganized — proactive as well as reactive approaches. Nokia, unlike Ericsson, had a plan in place for responding to financial disruptions caused by suppliers; it proactively increased procurement from its other suppliers (Tomlin, 2006).

6.2   Disruption Risk

Endogenous factors that increase the potential of supply chain disruptions include outsourcing, globalization, new technology, and reliance on sole suppliers As these factors also help build competitive advantage, the company needs to tradeoff increased profit with the diseconomies of disruption. For example, while outsourcing helps in cost reduction, it may suffer from a lack of control of the sources of production and distribution due to decentralized decision rights of partners. The decisions made by an outsourced partner may not be aligned with the company, resulting in supply failures and inventory build-up. Hazardous material from unreliable suppliers may have very disruptive impact on the supply chain. Another kind of disruption is that caused by failure of financial markets. The case in point is the mortgage loan crisis that erupted in the US in 2008/09 and engulfed the economy of the whole world in a deep recession. The linkages (or lack of) between the financial-securities of different countries created the avalanche effect.

Offshore outsourcing further increases the risk of disruption, as it involves suppliers who abide by the laws of their home countries. Global operations are also susceptible to currency fluctuations, and arbitrage. Long supply lines, spanning multiple countries, are "exposed" as sovereign nations may differ on their views on the adequate level of protection. Political, labor, and weather uncertainties add to the potential for disruption. Relief from disruptions can be slow because of slow moving government bureaucracies.

Governments may increase disruption potential by imposing new regulations. For example, national security concerns have caused the government of the United States to institute new terror related regulations in recent years. These include the C-TPAT (Customs-Trade Partnership Against Terrorism), AMR (Advanced Manifest Rule), CSI (Container Security Initiative), and SST (Smart and Secure Trade-lane). These laws, although designed to fend off major catastrophes, may help disrupt a company's daily operations. Both the C-TPAT, which provides for expedited processing at US ports (for certified shippers) and AMR, which requires detailed cargo data before shipment-arrival at US ports, are aimed at preventing dangerous shipments from landing at US ports. The certification process in C-TPAT introduces additional stages in the supply chain and can be onerous to small companies. Through bilateral agreements, CSI goes a step further by enabling the monitoring of packing and loading operations at the *outbound ports overseas*; SST requires electronic seals that can detect tampering of containers during transit to US ports. All such regulations have potential for disruptions through supply hold-backs and/or delays.

While a lean supply chain can create efficiency and reduce operations costs, it would not be an appropriate solution if a quick relief from disruptions is desired; lack of flexibility and redundancy in the supply chain may hinder agility. Similarly, while the practice of using sole suppliers for components builds efficiencies through scale economy and long term relationship, it also increases the risk from disruption because the supplier cannot be replaced quickly, if it went down. That is, fallback options are limited relative to the practice of using multiple suppliers per component.

6.3 Response Management

As mentioned earlier, events of disruption are typically characterized by a significant surge in demand for supplies, equipment, and manpower, depending on the likelihood and severity of the disruption. Historically, we note that likelihood of a disruption is inversely related to its severity.

Preparing for disruption implies a thorough analysis of the type and frequency of disruption, vulnerabilities in the supply chain, ownership of assets, estimates of consequences, available options and their prioritization, and how risks may be shared with partners. Response to disruption from terrorism would clearly be different from that from an environmental calamity, a rupture in financial markets, or operations failures. As all nodes and links are not equally vulnerable, preparation must include a vulnerability index and contingency plans for the most vulnerable nodes and links in the supply chain. Such contingency plans may include pre positioning of inventory and rerouting maps. They must be built upon estimates of damage from disruption and whether the vulnerable nodes and links are owned by the company or its partners. Such an analysis would reveal a set of recourse that may be available to the supply chain and how actions can be prioritized. Finally, a careful risk analysis must be undertaken to asses which risk types can be prevented, which can be mitigated, and which can be allocated to partners.

Managing response to disruptions comprises pure as well as hybrid approaches. The pure strategies are protection, proactive response, and reactive response; and hybrid strategies are combinations of two or more pure strategies. While protection is the containment of the negative impact of a disaster that leads to risk-reduction (Knemeyer et al., 2009) or decreased damage, reactive response is a real time response that bridges the gap between anticipated and actual disasters.

Proactive response determines the ex ante preparedness for disaster relief that mitigates risk. Pre positioning of inventory is based on risk management concepts. That is, inventory and capacity reserves are built and positioned strategically as hedges against the likelihood of disruptions. Kleindorfer and Saad (2005) and Lee (2004) discuss how backup systems, excess capacity, multiple suppliers, inventory buffers, and flexibility may lower the probability of disruptions as well as reduce its negative impact. In a similar vein, Chopra and Sodhi (2004) suggest that firms need to build reserves of inventory and capacity, and Tang (2006) emphasizes the need to build strategic stock of critical components.

Protection measures include fortification of production facilities (as in terrorism), strengthening of levees for flood control (as in hurricanes), and government regulations to fend off financial melt-down. Protection minimizes the need for disaster relief, though at high cost. As an example, the cost of rebuilding the New Orleans levees is estimated to be $10 billion (Whoriskey and Hsu, 2006).

6.3.1 Restructuring for Proactive Response

Both protection and pre positioning of inventories are examples of proactive response. There are several options for preventing disruptions. These include improving operations and building long term relationships with suppliers, designing robust products and processes, operating in less vulnerable niches, divesting from risky ventures, and gaining market power. Suppliers can be made more resilient by safeguarding their production and logistics operations, inventory holding, and tightly controlling all input materials. Transparency of supplier operations is the most effective way of thwarting impending disasters, as it can alert the partner before a supply failure. Transparency is possible with a long term relationship discussed in Chapter 5, where suppliers become "extended arms" of the enterprise. With expensive products, it becomes imperative to switch to modular design so that the damaged modules may be replaced quickly. Similarly, production and logistics processes should be designed with built-in flexibilities enabling rapid rerouting of products and transportation vehicles in case of a disruption. In terms of location, it would be prudent to relocate facilities away from disaster-prone zones, if feasible. This may also imply a reduction in offshore outsourcing. In the same vein, moving away from risky ventures such as global crude-oil refinery operations would make business sense. Clearly, implementation of the preventive steps become easier if the company occupies a dominant position in the market place.

The stakeholders in a supply chain may invest in a robust communication system, back up data and other assets, decentralize operations, and use multiple suppliers for risk diversification. If markets are susceptible to failures, suppliers could pool demand by supplying multiple customers from the same warehouse, and look to maintain customers from diverse markets. Where possible, suppliers would build in flexibility so that they are able to alter delivery plans after actual demands are known. With high demand volatility, forming alliances with other suppliers would minimize the risk of delivery-defaults. Advance purchase contracts help transferring demand risk from the supplier to the retailer. Similarly, outsourcing may transfer demand risk from the manufacturer to the outsourcing partners, if order size could be updated based on actual demand. A proactive strategy may, however, result in unutilized resources if disasters are infrequent.

In case of natural disasters the relief organizations and the national governments make advance plans for life-sustenance (shelter, food, medicine, blankets), support personnel, and equipment, all in anticipation. Customers may plan for time cushions so that if deliveries are disrupted they would have sufficient time to make alternative arrangements. This is especially important with natural disasters where life sustenance over a prolonged period is a critical factor.

6.3.2 Logistics for Reactive Response

A reactive strategy implies responding to the disaster in real time after it has occurred (ex post). The primary objective is saving human life. Thus, both quantity and speed of delivery are of utmost importance. As market forces are impaired, it would be hard to assess demand satisfactorily and, therefore, suppliers would need to fulfill demand with a significant information asymmetry. As supplies must be "pushed" to the disaster site, it would be akin to operating in a VMI mode but without the advantage of demand visibility. The most difficult problem to solve is logistics. There are several complicating issues. A logistics operation requires delivery trucks and vans, access from the supply sources (retailers) to disaster site by road network, adequate supplies, and a delivery plan specifying what needs to be delivered where. In the absence of any of the above four, the operation cannot be completed. As all disasters sites may not be reachable, one way would be to plan for strategic drop zones from where supplies can be air-dropped to the affected sites.

Information transparency in production, logistics, and fleet management is therefore an imperative. There is a dearth of rigorous research (conceptual or otherwise) on reactive approaches to disaster management. Qi et al. (2004) discusses a reactive strategy in a decentralized decision-making environment, and Tomlin (2006) explores dual-sourcing strategies with a single unreliable supplier using a discrete time Markov process.

6.3.3   The Hybrid Approach

The hybrid approach that integrates proactive and reactive modes uses two sources of inventory — pre positioned, and real time purchase. These decisions are impacted by two types of uncertainty: the severity of disruption, and assessment of the relief effort. While the severity level of the disaster would be known after it has occurred, the extent of needed relief may remain uncertain due to disruptions in information flow. Clearly, reactive relief efforts must be prepared for each severity level. In a decentralized decision structure (involving the relief provider, retailer, and the infrastructure provider), while the retailer and infrastructure providers are motivated by profit, the buyer (relief provider) would seek to minimize the social cost of resource denial and resource delays.

McEntire and Mathis (2005) and Mitchell (2003) outline how countries have evolved from using reactive to proactive approaches for disaster management. Mathematical models have been developed for integrating proactive capacity with reactive capacity. Cattani et al. (2008), and Eynan and Rosenblatt (1995), establish conditions for investment in both reactive and proactive capacities (also known as speculative capacity). Fisher and Raman (1996) discuss a capacity postponement model where an initial capacity is built based on demand forecast, with a provision for adding capacity later when demand becomes known.

6.3.4   Infrastructure

Damaged transportation routes and suppliers must be repaired or replaced. Replacing suppliers in real time can be difficult as it would imply incorporation of new suppliers with little known supply capabilities and locations. Although the utilization of resources in reactive response is high, the cost of procuring resources at a short notice and in a chaotic environment can also be high. For example, the cost of the initial response activity after the 9/11 attack including search and rescue, debris removal, emergency transportation, and emergency medical services, has been reported to be $2.55 billion. Demand uncertainties increase due to possible ruptures in the communication network that make demand estimates very unreliable. Disruption in communication also affects tracking, expediting, transactions, and supply chain coordination.

Most disruption relief efforts become stymied due to the damage to the infrastructure comprising communication, utilities, and transportation networks. Disruptions in telecommunication and utilities cause delays in relief supplies reaching their destination, as they often isolate entire regions. Repair of utility lines and cables can be slow, especially, if the ruptures are widespread. One way of motivating the service providers is to provide incentives for them to build robust networks, and carry out preventive maintenance (proactively). Disruptions in the transportation network can further delay the deliveries to the sites, as the roads and bridges must be repaired. Note that delays cause losses in business operations. In natural disasters, they also cause hardships, increasing the social cost of the disaster. Two major issues would, therefore, be of considerable interest: (i) how to construct and evaluate hybrid relief strategies that include both proactive and reactive responses, and (ii) how to induce collaboration in the network that would benefit all.

6.3.5   Risk management

Managing risk in a supply chain, especially, risk mitigation is discussed in Chopra and Sodhi (2004), and in Hendricks and Singhal (2005). Tomlin (2006), Swinney and Netessine (2009), and others (Albeniz and Simchi-Levi, 2005; Arreola-Risa and DeCroix, 1998a) have investigated effectiveness of various mitigation strategies. Supply chain's vulnerability to disasters is studied by Stauffer (2003), and the dimension of qualitative measures of vulnerabilities is explored by Svensson (2004). Zsidisin et al. (2005) discusses supply risk and outlines a framework with four stages — awareness, prevention, remediation, and knowledge management. Song et al. (2000) discuss uncertainties in supply and demand in the context of assembly contracts. Stecke and Kumar (2009) classify mitigation strategies as proactive, advance-warning, and coping, and show how they relate to vulnerabilities in practices such as globalization, outsourcing, decentralization, and JIT. Assessing the

Situation	Risk allocation	How
Supplier assets not specific to the customer	Allocated to supplier	Supplier takes responsibility of investment decisions
Non strategic relationship	Allocated to supplier	Supplier guarantees a minimum availability
High asset intensity	Allocated to customer	Customer owns technology
High profit uncertainty	Allocated to customer	Advance-purchase by customers
Strategically important major customer	Allocated to customer	Customer guarantees minimum purchases
High asset intensity	Risk sharing	Long term contracts
Collaboration	Risk sharing	Profit sharing

Fig. 6.1 Risk allocation.

probability of rare events (such as the 9–11 terrorist attack) has been mostly conceptual in nature to date. Sheffi and Rice (2005) discusses procedures for estimating the likelihood of rare events in the context of General Motors' supply chain. They discuss how supply chains can be made robust and resilient through flexibility in supply, procurement, distribution, and organizational culture. Craighead et al. (2007) suggests that the impact of disruptions is inversely related to the presence of relief capabilities.

One of the major issues in risk management in a supply chain is establishing who would absorb how much of the disruption risk. For supply chain effectiveness risk must be allocated to the party that controls the risky operations. A few examples illustrating these concepts are shown in Figure 6.1. Note that a feasible solution would be to allocate risk to the party that has the resources to bear it most.

6.4 Defensive Approaches

Today security from global terrorism has become central to disruption management in global supply chains, as the traditional financial and operational response strategies are less effective against it. Terrorists may attack facilities owned by the company or they may attack a supplier's facility. In both cases the supply chain is damaged, but with different consequences.

Consider first the case of facilities owned by a company. As the company cannot know with certainty which of its facilities ($i \in I$) will be attacked, its best strategy would be to protect only a subset ($I_1 \subseteq I$) of facilities. The attacker may want to attack the undefended facilities, but not all undefended facilities would be high-valued targets. We assume that facility i generates a revenue p_i for the company.

Consider the scenario where the attacker decides to attack facilities in the set I_2 ($I_2 \subseteq I$). It follows that the attacker would determine best I_2 ($= I_2^*$) for a known I_1, and then the company would determine the best I_1, corresponding to the attacker best response I_2^*. We let the binary decision variables R_j and L_i denote whether facility j is attacked and facility i is defended, respectively. We stipulate $R_j = 1$ if there is an attack on facility j (0 otherwise), and $L_i = 1$ if facility i is defended (0 otherwise). We assume that if an undefended facility is attacked it would be destroyed completely, so that the value to the attacker of destroying an undefended facility i would be p_i. If the facility is defended, the damage inflicted by an attack decreases and, therefore, the value to the attacker would be reduced by v_i. The cost of defending and attacking resource are assumed to be $a(I_1)$ and $b(I_2)$ respectively.

For a known I_1 (i.e. known vector L), the attacker's value in attacking the facilities would be $\sum_j (p_j - v_j L_j) R_j - b(I_2)$. Hence, the attacker's objective function would be

$$\operatorname*{Max}_{I_2 \subseteq I} \left\{ \sum_j (p_j - v_j L_j) R_j - b(I_2) \right\}, \quad \text{given } L_{j=1}, j \in I_1$$

The attacker's optimal decision would be I_2^* (i.e. R^*).

The company, as the leader, would maximize her objective function in I_1 (i.e. L_i).

$$\operatorname*{Max}_{I_1 \subseteq I} \left[\sum_i \sum_j \{p_i (1 - R_j^*) + v_j L_i R_j^*\} - a(I_1) \right]$$

Using the solution to the above problem, we can determine the criticality of a group of facilities, i.e., the value of protecting a given set of assets. We can also determine the value of adding additional facilities to improve the system's robustness.

Next, consider the scenario where a supplier's facility is attacked. Effective strategies for security in supply chains need to encompass the entire supply chain of all participating organizations. If a supplier is damaged, the losses incurred by the linked supply chain might exceed by far the value derived from that individual supplier. Consider a simple supply chain with a retailer and a supplier who is susceptible to disruption from terrorism. If the supplier is hit, it loses its entire capacity. In which case, we assume, that a second supplier is available at a much higher cost. The retailer or the supplier can, of course, enhance protection of the supply chain at a cost. We denote r as the probability that supply chain is damaged in an attack. The higher the investment in protection the lower will be the damage-probability.

We consider two cases: the retailer invests X_R or the supplier invests X_S to protect the supply link between them. First, when the retailer invests in supply chain protection, we express the damage probability as $r = e^{-X_R}$, as there is a diminishing return to investment.

The retailer's profit is expressed as,

$$
\pi_R = \left| \begin{array}{l} (1-r)\left[p\left\{ \int_{\xi \leq q} \xi f(\xi)d\xi + q\int_{\xi \geq q} f(\xi)d\xi \right\} - wq \right] \\ \\ + r\left[p\left\{ \int_{\xi \leq Q} \xi f(\xi)d\xi + Q\int_{\xi \geq Q} f(\xi)d\xi \right\} - WQ \right] - X_R \end{array} \right| \tag{6.1}
$$

In (6.1), sales price is assumed to be p, the wholesale prices of the suppliers are w and W, and purchase quantities are q and Q.

The suppliers' profit are expressed as,

$$
\pi_S = (1-r)(w-c)q \tag{6.2}
$$

$$
\pi_{S2} = r(W-C)Q \tag{6.2a}
$$

From the first order conditions in X_R, w, and q, as shown in Appendix I, are

$$e^{X_R} = p\left\{\int_{\xi \leq q} \xi f(\xi)d\xi + q\int_{\xi \geq q} f(\xi)d\xi\right\}$$

$$-p\left\{\int_{\xi \leq Q} \xi f(\xi)d\xi + Q\int_{\xi \geq Q} f(\xi)d\xi\right\} + WQ - wq \quad (6.3)$$

$$w = p\{1 - F(q)\} \quad (6.4)$$

$$1 - F(q) - qf(q) + \frac{e^{-2X_R}}{1 - e^{-X_R}}pq^2 f(q)\left\{1 - F(q) - \frac{c}{p}\right\} = \frac{c}{p} \quad (6.5)$$

To obtain optimal Q and W, we generate two additional equations similar to equations (6.4) and (6.5). Thus,

$$W = p\{1 - F(Q)\} \quad (6.4a)$$

$$1 - F(Q) - Qf(Q) + \frac{e^{-2X_R}}{1 - e^{-X_R}}pQ^2 f(Q)\left\{1 - F(Q) - \frac{C}{p}\right\} = \frac{C}{p} \quad (6.5a)$$

$$C > c$$

Solve (6.3), (6.4), (6.4a), (6.5) and (6.5a) for q, w, Q, W, and X_R.
    Rewriting (6.5)

$$1 - F(q) - qf(q) = \frac{c}{p} - \left(\frac{1}{e^{2X_R} - e^{X_R}}\right)\left\{1 - F(q) - \frac{c}{p}\right\}pq^2 f(q)$$

Note that $1 - F(q) < \frac{c}{p} \Rightarrow 1 - F(q) - qf(q) < \frac{c}{p}$, which is not possible as the RHS would exceed c/p. Therefore, as $1 - F(q) > \frac{c}{p}$, the RHS of the above equation increases in X_R for a given q. The *only way* the equation can remain balanced is for the LHS to increase (though not fully as the RHS decreases in q). Note that the LHS decreases in q (IGFR property). Therefore, the optimal q must decrease in X_R. This implies that *as the supply link becomes more secure, the buyer becomes less concerned about disruption and reduces the order size* (anti hoarding mindset).
    Next substituting from (6.4) and (6.4a) into (6.1), we obtain

$$e^{X_R} = p\left\{\int_{\xi \leq q} \xi f(\xi)d\xi - \int_{\xi \leq Q} \xi f(\xi)d\xi\right\}$$

Thus, the presence of the 2nd supplier, helps decrease investment in supply chain protection. Next, when the supplier invests X_S in supply chain protection, we would have $r = e^{-X_S}$.

The retailer's profit is expressed as,

$$\pi_R = \begin{vmatrix} (1-r)\left[p\left\{\int_{\xi \le q}\xi f(\xi)d\xi + q\int_{\xi \ge q}f(\xi)d\xi\right\} - wq\right] \\ +r\left[p\left\{\int_{\xi \le Q}\xi f(\xi)d\xi + Q\int_{\xi \ge Q}f(\xi)d\xi\right\} - WQ\right] \end{vmatrix} \tag{6.6}$$

The supplier's profit is expressed as,

$$\pi_S = (1-r)(w-c)q - X_S \tag{6.7}$$

Hence, from the first order condition, established in Appendix II,

$$w = p\{1 - F(q)\} \tag{6.8}$$

$$1 - F(q) - qf(q) = \frac{c}{p} \tag{6.9}$$

$$e^{X_S} = (w-c)q \tag{6.10}$$

Solve (6.8), (6.9), and (6.10) for q, w, and X_S.

To solve for the optimal Q and W, we can generate two additional equations as before.

6.5 Natural Disaster Relief

Consider the supply chain comprising a government agency such as FEMA or a non-profit organization, and two service providers: a retailer of relief supplies, and an infrastructure provider. The buyer decides how much relief-supply to purchase and the retailer determines at what price to sell. The infrastructure provider maximizes his profit by appropriately deciding the repair time for infrastructure repair, and the amount of investment for increasing the robustness of the infrastructure. His revenues would of course be linked to the repair time he achieves.

Disaster relief management processes from the government sources outline systems and procedures for coordinating different relief

Hybrid Strategies				
Pure Strategies	H1	H2	H3	H4
Protection	x		x	x
Proactive Response	x	x		x
Reactive Response	x	x	x	

Fig. 6.2 Hybrid strategies.

agencies. During a disaster incident, numerous procedures and administrative functions are required for support. FEMA (2009) describes the roles and responsibilities of federal departments, local entities, the private sector, and volunteer organizations; and how they coordinate and execute common processes and administrative tasks for efficient management of disasters. Chakravarty (2005b) discusses an online platform for collaboration between government agencies and non-profit organization, involved in disaster relief.

Note that providing complete protection can be very costly, if not impossible. Real time acquisition of supplies can also be expensive. Therefore, tradeoffs would exist between the pure strategies of protection, proactive response, and reactive response. This suggests that a hybrid strategy may perform better. In Figure 6.2, combinations of two out of three pure strategies, to form four hybrid strategies, H1 to H4, are shown.

It seems natural to characterize hybrid strategies based on disaster-severity level, as it would be known in real time after the disruption has occurred. In disasters such as hurricanes, the wind speed and hence the severity of the hurricane is known. We characterize a hybrid strategy by two parameters: t and τ, both are threshold values for the severity level. A rank-ordering of severity level η $(0 \leq \eta \leq \infty)$ is often used. Hurricanes, for example, are categorized into five categories: category 1 the least severe and category 5 the most severe. In Figure 6.2, Hybrid H1 implies that protection is built-up to shield against disasters up to severity level $\eta = t$, proactive acquisition of supplies for $t \leq \eta \leq \tau$, and real time purchase of supplies for $\eta \geq \tau$. Note that H2 implies $t = 0$, H3 implies $t = \tau$, and H4 implies $\tau = \infty$.

One of the difficult issues in natural disaster management is assessing the amount of relief supplies needed at a disaster site. The buyer (government agency or non-profit organization) estimates the need based on field information. This assessment (for a known η), is uncertain at best and can be expressed as a random variable ξ_η with distribution $F_\eta(\cdot)$. Note that ξ_η is generic demand for supplies, easily converted to actual needs such as medicine, clothing, food, shelter, and support personnel. We assume ξ_η to be stochastically increasing in η. Thus, $E(\xi_\eta) = \mu_\eta$ increases in η, skewing the distribution to the right.

The buyer contracts for relief-supplies q at a unit-cost w. We use q, w, and q_η, w_η for the proactive and reactive scenarios, respectively. A supply shortage ($\xi_\eta - q$) implies possible loss of life and hence a high social cost. Denote this cost as $r(r > w)$ per unit. We assume that the excess inventory ($q - \xi_\eta$) can be salvaged by the buyer at a price s per unit.

6.6   A Hybrid Strategy for Natural Disasters

Delays in delivering supplies to the affected site cause hardships to the victims and help transform the victim's hardship into buyer's social cost. Delays are primarily caused by damages to infrastructure (transportation and communication networks). One way to minimize delay is to either build a robust infrastructure, or carry out preventive maintenance and rapid repairs of ruptures. Note that demand fulfilment time includes a transportation time, in addition to the infrastructure repair time. In a proactive mode the buyer may preposition the supplies, and thus reduce transportation time to the disaster site.

The contract between the buyer and the infrastructure service provider, to maintain and repair the network, can be designed to provide incentives to keep the repair times low. Buyer may agree to pay an amount dependent on the repair time that the service provider can achieve. The service provider establishes the best value of repair time based on his cost structure and the buyer decides the amount to be paid in response to the repair time.

Consider the hybrid strategy H2 for a given τ, which includes proactive and reactive responses but no risk reduction through protection (that is, $t = 0$). It follows that there will be real time purchase of supplies only if $\eta \geq \tau$. The buyer contracts a quantity q before the disruption and q_η after the disruption. In stage 1, decisions are made for q and q_η for known values of w and w_η. In stage 2, decisions are made for w and w_η by the retailers, using stage 1 decisions as reaction functions in a Stackelberg game.

Chakravarty (2009a) establishes that for any τ, the required amount of relief supplies has two components, q^* and q_η^*. Note that q^* increases (at a decreasing rate) in τ; it never exceeds q_p, the quantity acquired in a pure proactive mode. Note that q_η^* is zero for $0 \leq \eta \leq \tau$. For a given η, q_η^* decreases in τ, $(\tau \leq \eta)$. Therefore, if the disaster is not severe both parties settle for the lower quantity (q^*). In severe disaster scenarios, an additional quantity of q_η^* is acquired, and the total purchase quantity in the hybrid mode is at least as large as q_p. Thus, at $\eta = \tau$, there is a step-jump as the total purchase quantity increases from q^* to $q^* + q_\eta^* \geq q_p$. When $\tau \to \infty$, $q^* + q_\eta^* \to q_p$.

Note that a reduction in infrastructure- repair time decreases the social cost of delivery delays, which motivates acquisition of additional supplies bringing more revenue to retailer. The savings from shorter delays in delivery are thus shared three ways: to reward the infrastructure provider for faster repairs, to increase the retailer's profit through higher purchase quantity of supplies, and to reduce the buyer's total cost. Of course, a shorter repair time also decreases the hardship caused by delays at the disaster site. This is win-win for all.

6.7   Appendix I

Optimizing π_R in X_R and q, for given Q and W.

$$\frac{\partial \pi_R}{\partial X_R} = e^{-X_R}\left[p\left\{\int_{\xi \leq q} \xi f(\xi)d\xi + q\int_{\xi \geq q} f(\xi)d\xi\right\}\right.$$
$$\left. - p\left\{\int_{\xi \leq Q} \xi f(\xi)d\xi + Q\int_{\xi \geq Q} f(\xi)d\xi\right\} + WQ - wq\right] - 1$$

$$\frac{\partial \pi_R}{\partial q} = (1 - e^{-X_R})[p\{1 - F(q)\} - w]$$

As π_R is concave in X_R and q, we have from first order conditions

$$e^{X_R} = p\left\{\int_{\xi\leq q}\xi f(\xi)d\xi + q\int_{\xi\geq q}f(\xi)d\xi\right\}$$

$$-p\left\{\int_{\xi\leq Q}\xi f(\xi)d\xi + Q\int_{\xi\geq Q}f(\xi)d\xi\right\} + WQ - wq$$

$$w = p\{1 - F(q)\}$$

Optimizing π_S in q (proxy for w),

$$\frac{\partial\pi_S}{\partial q} = (1-r)\left\{q\frac{\partial w}{\partial q} + (w-c)\right\} - \frac{\partial r}{\partial q}(w-c)q$$

Note that from (6.3) and (6.4) we obtain

$$\frac{\partial w}{\partial q} = -pf(q), \quad \text{and} \quad \frac{\partial r}{\partial q} = \frac{\partial r}{\partial X_R}\cdot\frac{\partial X_R}{\partial q}$$

$$= -e^{-2X_R}\left[p\{1 - F(q)\} - w - q\frac{\partial w}{\partial q}\right]$$

Note that

$$\frac{\partial X_R}{\partial q} = e^{-X_R}\left[p\{1-F(q)\} - w - q\frac{\partial w}{\partial q}\right], \frac{\partial r}{\partial X_R} = -e^{-X_R}.$$

Substituting for $\frac{\partial w}{\partial q}$ and w, $\frac{\partial r}{\partial q}$ simplifies to $\frac{\partial r}{\partial q} = -e^{-2X_R}qpf(q)$.
    Substituting for $\frac{\partial w}{\partial q}$, w, and $\frac{\partial r}{\partial q}$ in $\frac{\partial\pi_S}{\partial q}$ we have,

$$\frac{\partial\pi_S}{\partial q} = (1 - e^{-X_R})p\{1 - F(q) - qf(q)\}$$

$$+e^{-2X_R}p^2q^2f(q)\left\{1 - F(q) - \frac{c}{p}\right\} - c(1 - e^{-X_R})$$

Assuming IGFR property for $F(\cdot)$, the first term clearly decreases from a positive to a negative value in q. In the second term, the expression inside the parenthesis, $1 - F(q) - \frac{c}{p}$, decreases from a positive to a negative value in q (as $p > c$). Hence, π_S is unimodal in q, and we have from the first order condition

$$1 - F(q) - qf(q) + \frac{e^{-2X_R}}{1 - e^{-X_R}}pq^2f(q)\left\{1 - F(q) - \frac{c}{p}\right\} = \frac{c}{p}$$

6.8   Appendix II

Optimizing π_R in q, for given Q and W.

$$\frac{\partial \pi_R}{\partial q} = (1 - e^{-X_S})[p\{1 - F(q)\} - w]$$

As π_R is concave in q,

$$w = p\{1 - F(q)\}$$

Optimizing π_S in q (proxy for w),

$$\frac{\partial \pi_S}{\partial q} = (1 - r)\left\{ q\frac{\partial w}{\partial q} + (w - c) \right\}$$

Note that from (6.8) we obtain $\frac{\partial w}{\partial q} = -pf(q)$.

Substituting for $\frac{\partial w}{\partial q}$ and w in $\frac{\partial \pi_S}{\partial q}$ we have,

$$\frac{\partial \pi_S}{\partial q} = (1 - e^{-X_S})p\{1 - F(q) - qf(q)\} - c(1 - e^{-X_S})$$

Assuming IGFR property for $F(\cdot)$, the first term clearly decreases from a positive to a negative value in q. Hence, π_S is unimodal in q, and from the first order condition,

$$1 - F(q) - qf(q) = \frac{c}{p}$$

Next, the supplier can determine the optimal investment as shown below

$$\frac{\partial \pi_S}{\partial X_S} = e^{-X_S}(w - c)q - 1$$

As π_S is concave in X_S, from the first order condition,

$$e^{X_S} = (w - c)q$$

Acknowledgments

This book is made possible with the support of many individuals and organizations. Foremost is the Fulbright Commission, who got the ball rolling by awarding me a distinguished chair in supply chain at the Vienna University of Economics and Business. I am thankful for the opportunity to "experiment" with emerging concepts on relevant topics at doctoral seminar classes at the VU. The doctoral class of 2008 was superb as a sounding board and it helped shaped many of my ideas in their early stages. Many thanks to Alfred Taudes, the department chair, for facilitating it.

I am thankful for the support of the College of Business Administration at Northeastern University through the Philip McDonald Chair endowment. The president of Northeastern University and the dean of the College of Business Administration deserve special mention. I am fortunate to have worked with many students and colleagues, who contributed with valuable insights.

This book could not have been completed without the continuous support of my family. I remain indebted forever.

References

Albeniz, A. and D. Simchi-Levi (2005), 'A portfolio approach for procurement contracts'. *Production and Operations Management* **14**(1), 90–114.

Arreola-Risa, A. and G. DeCroix (1998a), 'Inventory management under random supply disruptions and partial backorders'. *Naval Research Logistics* **45**, 687–703.

Arreola-Risa, A. and G. DeCroix (1998b), 'Make-to-order versus make-to-stock in a production inventory system with general production times'. *IIE Transactions* **30**, 705–713.

Aviv, Y. (2001), 'The effect of collaborative forecasting on supply chain performance'. *Management Science* **47**(10), 1326–1343.

Baldwin, C. and K. Clark (1997), 'Managing in an age of modularity'. *Harvard Business Review* pp. 84–93.

Basu, A. and A. Kumar (2002), 'Research commentary: Workflow management issues in e-business'. *Information Systems Research* **13**(1), 1–14.

Beamon, M. (1999), 'Designing the green supply chain'. *Logistics Information Management* **12**, 332–342.

Benjaafar, S., M. ElHafsi, and F. Véricourt (2004), 'Demand allocation in multiple-product, multiple-facility, make-to-stock systems'. *Management Science* **50**(10), 1431–1448.

Bitran, G. and R. Caldentey (2003), 'An overview of pricing models for revenue management'. *Manufacturing & Service Operations Management* **5**(3), 203–229.

Bolton, P. and M. Dewatripoint (2004), *Contract Theory*. Cambridge, Massachusetts: The MIT Press.

Boute, R., S. Disney, M. Lambrecht, and B.VanHoudt (2009), 'Designing replenishment rules in a two-echelon supply chain with a flexible or an inflexible capacity strategy'. *International Journal of Production Economics* **119**, 187–198.

Bursa, K. (2009), 'How to effectively manage demand with demand sensing and shaping using point of sales data'. *Journal of Business Forecasting* pp. 26–28.

Cachon, G. (2003), 'Supply chain coordination with contracts'. In: S. Graves and T. De Kok (eds.): *Invited chapter in Handbooks in Operations Research and Management Science, 11, Supply Chain Management: Design, Coordination and Operation*. North Holland, Amsterdam, pp. 229–339.

Cattani, D., E. Dahanb, and G. Schmidt (2008), 'Tailored capacity: Speculative and reactive fabrication of fashion goods'. *International Journal of Production Economics* **114**, 416–430.

Chakravarty, A. (1997), 'A model for switching dispatching rules in real-time in a flexible manufacturing cell'. *Production and Operations Management* **6**, 398–418.

Chakravarty, A. (2005a), 'Global plant capacity and product allocation with pricing'. *European Journal of Operational Research* **165**, 157–181.

Chakravarty, A. (2005b), 'Tsunami relief coordination'. Northeastern University, College of Business Administration, http://w.cba.neu.edu:80/~akc/amiya.html.

Chakravarty, A. (2009a), 'Disaster disruption relief: A supply chain response'. Working paper, CBA, Northeastern University, Boston, MA.

Chakravarty, A. (2009b), 'Offshore business process outsourcing: The build-operate-transfer model'. Working paper, CBA, Northeastern University, Boston, USA.

Chakravarty, A. and N. Blakrishnan (2001), 'Achieving product variety through optimal choice of module variations'. *IIE Transactions* **33**, 587–598.

Chakravarty, A. and N. Blakrishnan (2008), 'Product design with multiple suppliers for component variants'. *International Journal of Production Economics* **112**, 723–741.

Chakravarty, A. and S. Ghose (1993), 'Tracking product-process interactions: A research paradigm'. *Production and Operations Management* **2**(2), 72–93. Spring.

Chakravarty, A., A. Mild, and A. Taudes (2010), 'Bundling decisions in a supply chain'. Working paper, College of Business Administration, Northeastern University, Boston, USA.

Chakravarty, A. and J. Zhang (2007), 'Lateral capacity exchange and its impact on capacity investment decisions'. *Naval Research Logistics* **54**, 632–644.

Chandran, M. and V. Gupta (2003), 'Wal-Mart's supply chain management practices'. *ICFAI Management Redsearch*. http:// mohanchandran.files.wordpress.com/2008/01/wal-mart.pdf.

Chopra, S. and M. Sodhi (2004), 'Managing risk to avoid supply chain breakdown'. *MIT Sloan Management Review* **46**(1), 53–61.

Clark, T. and J. McKenney (1994), 'Campbell soup company: A leader in continuous replenishment innovations'. HBS Case 9-195-124, Harvard University.

Coughlan, P., M. Rukstad, and C. Johnston (2001), *WebMD (A), HBS Case Study, 701007-PDF-ENG*. Cambridge, MA: Harvard University.

Craighead, C., J. Blackhurst, M. Rungtusanatham, and R. Handfield (2007), 'The severity of supply chain disruptions: Design characteristics and mitigation capabilities'. *Decision Sciences* **38**(1), 131–156.

Dapiran, P. (1992), 'Benetton — global logistics in action'. *Asia-Pacific International Journal of Business Logistics* **5**, 7–11.

Eynan, A. and M. Rosenblatt (1995), 'Assemble to order and assemble in advance in a single-period stochastic environment'. *Naval Research Logistics* **42**, 861–872.

FEMA (2009), 'Emergency support function annex'. http://www.fema. gov/pdf/emergency/nrf/nrf-support-intro.pdf.

Fine, C. (1998), *Clockspeed: Winning Industry Control in the Age of Temporary Advantage*. Perseus Books, Reading, MA.

Finkle, J. (2005), 'Economic Development after Katrina: Lessons from National Post-Disaster Response and Recovery'. http://www. iedconline.org/Downloads/GulfCoast/Katrina_ Lessons.pdf.

Fisher, M. and A. Raman (1996), 'Reducing the cost of demand uncertainty through accurate response to early sales'. *Operations Research* **44**(1), 87–99.

Gerchak, Y. and H. Q (2002), 'On the relation between the benefits of risk pooling and the variability of demand'. Working Paper, Tel Aviv University, Israel.

Gerwin, D. (1993), 'Manufacturing flexibility: a strategic perspective'. *Management Science* **39**(4), 395–410.

Ghemawat, P. and J. Nueno (2003), 'Zara: Fast fashions'. HBS Case 9-703-497.

Ghosn (2009), 'Ten years into the Alliance, Renault and Nissan are taking cooperation to a higher level'. http://www.ameinfo. com/198894.html.

Hartman, E., T. Ritter, and H. Gemuenden (2001), 'Determining the Purchase Situation: Cornerstone of Supplier Relationship Management'. 17th Annual IMP Conference, Norwegian School of Management BI, Oslo, Norway, 9th–11th September.

Hauser, J. (1993), 'How puritan-bennett used the house of quality'. *Sloan Management Review*, Spring pp. 61–68.

Hauser, J. and D. Clausing (1988), 'The house of quality'. *Harvard Business Review* pp. 63–73.

Hendricks, K. and V. Singhal (2005), 'An empirical analysis of the effect of supply chain disruptions on long-run stock price performance and equity risk of the firm'. *Production and Operations Management* **14**(1), 35–52.

Himmelman, A. (1993), 'Helping each other help others: Principles and practices of collaboration'. ARCH Fact Sheet Number 25, National Resource Center for Respite and Crisis Care Services: Chicago http://www.chtop.com/ARCH/archfs25.htm.

Ho, M. (2006), *Dell: Overcoming Roadblocks to Growth, HBS Cases, HKU 575*. University of Hong Kong: Asia Case Research Center.

Jaikumar, R. (1986), 'Postindustrial manufacturing'. *Harvard Business Review* **64**(6), 69–76.

Jinhong, X. and S. Shugan (2001), 'Electronic tickets, smart cards, and online prepayments: When and how to advance sell'. *Marketing Science* **20**(3), 219–243.

Jordan, W. and S. Graves (1995), 'Principles on the benefits of manufacturing process flexibility'. *Management Science* **41**, 577–594.

Kalakota, R. and M. Robinson (2004), *Offshore Outsourcing: Will Your Job Disappear in 2004*. Addison Wesley.

Kleindorfer, P. and G. Saad (2005), 'Managing disruption risks in supply chains'. *Production and Operations Management* **14**(1), 53–68.

Knemeyer, A., W. Zinn, and C. Eroglu (2009), 'Proactive planning for catastrophic events in supply chains'. *Journal of Operations Management* **27**, 141–153.

Kraljic, P. (1983), 'Purchasing must become supply management'. *Harvard Business Review* **61**, 109–117.

Kuzgunkaya, O. and H. ElMaraghy (2007), 'Economic and strategic perspectives on investing in RMS and FMS'. *International Journal of Flexible Manufacturing System* **19**, 217–246.

Lau, A., R. Yam, and E. Tang (2007), 'The impacts of product modularity on competitive capabilities and performance: An empirical study'. *International Journal of Production Economics* **105**, 1–20.

Lee, C. (2003), 'Demand chain optimization: pitfalls and key principles'. Evant White Paper Series.

Lee, H. (2002), 'Aligning supply chain strategies with product uncertainties'. *California Management Review* **44**(3).

Lee, H. (2004), 'Supply chain security — Are you ready?, SGSCMF-W1-2004'. Stanford Global Supply Chain Management Forum, Stanford University.

Lee, H., P. Padmanabhan, and S. Whang (1997), 'Information distortion in a supply chain: The bullwhip effect'. *Management Science* **43**(4), 546–558.

Lee, H. and S. Whang (2001), 'E-business and supply chain integration'. SBSCMB-W2-2001, Stanford Global Supply Chain Management Forum, Stanford University, Palo Alto, California, USA.

Lee, H. and S. Whang (2002), 'The impact of the secondary market on the supply chain'. *Management Science* **48**, 719–731.

Linden, G., B. Smith, and J. York (2003), 'Amazon.com Recommendations: Item-to-Item collaborative filtering'. *IEEE Internet Computing* pp. 76–80.

McEntire, D. and S. Mathis (2005), 'Comparative politics and disasters: Assessing substantive and methodological contributions'. Department of Public Administration, University of North Texas, http://training.fema.gov/EMIweb/edu/docs/EMT/Chapter%20-%20 Comparative%20Politics%20and%20Disasters.doc.

Mcgill, J. and G. Van Ryzin (1999), 'Revenue Management: Research Overview and Prospects'. *Transportation Science* **33**, 233–256.

Metters, R. (2008), 'A typology of offshoring and outsourcing in electronically transmitted services'. *Journal of Operations Management* **26**, 198–211.

Mitchell, T. (2003), 'An operational framework for mainstreaming disaster risk reduction'. Department of Geography, University College London, 26 Bedford Way, London, http://www.abuhrc. org/Publications/Working%20Paper%208.pdf.

Murakoshi, T. (1994), 'Customer driven manufacturing in Japan'. *International Journal of Production Economics* **37**, 63–72.

Page, A. and H. Rosenbaum (1987), 'Redesigning product lines with conjoint analysis: How sunbeam does it'. *Journal of Product Innovation Management* **4**, 120–137.

Qi, X., J. Bard, and G. Yu (2004), 'Supply chain coordination with demand disruptions'. *OMEGA: International Journal of Management Science* **32**, 301–312.

Rajagopalan, S. (2002), 'Make-to-order or make-to-stock: Model and application'. *Management Science* **48**, 241–256.

Rice, J. and F. Caniato (2003), 'Building a secure and resilient supply network'. *Supply Chain Management Review* **7**(5), 22–30.

Rudi, N., S. Kapur, and D. Pyke (2001), 'A two-location inventory model with transshipment and local decision making'. *Management Science* **47**, 1668–1680.

Seifert, R. (2003), "mi adidas" *Mass Customization Initiative, HBS Case Study, IMD159-PDF-ENG*. Cambridge, MA: Harvard University.

Sheffi, Y. and J. Rice (2005), 'A supply chain view of the resilient enterprise'. *MIT Sloan Management Review* **47**(1), 41–48.

Song, J., C. Yano, and P. Lerssrisuriya (2000), 'Contract assembly: Dealing with combined supply lead time and demand quantity uncertainty'. *Manufacturing and Service Operations Management* **2**(3), 287–296.

Stauffer, D. (2003), 'Risk: The weak link in your supply chain'. *Harvard Management Update* **8**(3), 3–5.

Stecke, K. and S. Kumar (2009), 'Sources of supply chain disruptions, factors that breed vulnerability, and mitigating strategies'. *Journal of Marketing Channels* **16**(3), 193–226.

Stratton, R. and R. Warburton (2003), 'The strategic integration of agile and lean supply'. *International Journal of Production Economics* **85**, 183–198.

Stremersch, S. and G. Tellis (2002), 'Strategic bundling of products and prices: A new synthesis for marketing'. *Journal of Marketing* **66**(1), 55–72.

Swinney, R. and S. Netessine (2009), 'Long-term Contracts under the Threat of Supplier Default'. *Manufacturing & Services Operations Management* **11**(1), 109–127.

Talluri, K. and G. van Ryzin (2004), *Theory and Practice of Revenue Management*. New York: Kluwer Academic Publishers.

Tang, C. (2006), 'Perspectives in supply chain risk management'. *International Journal of Production Economics* **103**(2), 451–488.

Tomlin, B. (2006), 'On the value of mitigation and contingency strategies for managing supply chain disruption risks'. *Management Science* **52**, 639–657.

VanMieghem, J. (2008), *Operations Strategy, Principles and Practice, Dynamics Ideas.* Belmont, Mass, USA.

VICS (2001), 'CPFR: Collaboration Planning, Forecasting, and Replenishment'. www.cpfr.org.

Waller, B. (2004), 'Market responsive manufacturing for the automotive supply chain'. *Journal of Manufacturing Technology Management* **15**, 10–19.

Warburton, R. and R. Stratton (2002), 'Questioning the relentless shift to offshore manufacturing'. *Supply Chain Management* **7**, 101–108.

Weatherford, L. (1997), 'Optimization of joint pricing and allocation perishable-asset revenue management problems with cross-elasticity'. *Journal of Combinatorial Optimization* **1**, 277–304.

Whoriskey, P. and S. Hsu (2006), 'Levee Repair Costs Triple'. Washington Post, March 31, pp A01, http://www.washingtonpost.com/wp-dyn/content/article/2006/03/30/AR2006033001912.html.

Zapfel, G. and M. Wasner (2002), 'Planning and optimization of hub-and-spoke transportation networks of cooperative third-party logistics providers'. *International Journal of Production Economics* **27**, 207–220.

Zipkin, P. (2001), 'The limits of mass customization'. *Sloan Management Review* **42**.

Zsidisin, G., L. Ellram, J. Carter, and J. Cavinato (2004), 'An analysis of supply risk assessment techniques'. *International Journal of Physical Distribution* **34**, 397–413.

www.ingramcontent.com/pod-product-compliance
Lightning Source LLC
Chambersburg PA
CBHW061333220326
41599CB00026B/5158